Ecology and the Environment: Perspectives from the Humanities

Religions of the World and Ecology Series

Previous volumes in the Religions of the World and Ecology series, for which Mary Evelyn Tucker and John Grimm were series editors, are available through Harvard University Press:

- *Buddhism and Ecology: The Interconnection of Dharma and Deeds,* Mary Evelyn Tucker and Duncan Ryûken Williams, eds.
- *Christianity and Ecology: Seeking the Well-Being of Earth and Humans,* Dieter T. Hessel and Rosemary Radford Ruether, eds.
- *Confucianism and Ecology: The Interrelation of Heaven, Earth, and Humans,* Mary Evelyn Tucker and John Berthrong, eds.
- *Daoism and Ecology: Ways within a Cosmic Landscape,* N. J. Girardot, James Miller, and Liu Xiaogan, eds.
- *Hinduism and Ecology: The Intersection of Earth, Sky, and Water,* Christopher Key Chapple and Mary Evelyn Tucker, eds.
- *Indigenous Traditions and Ecology: The Interbeing of Cosmology and Community,* John A. Grimm, ed.
- *Islam and Ecology: A Bestowed Trust,* Richard C. Foltz, Frederick M. Denny, and Azizan Baharuddin, eds..
- *Jainism and Ecology: Nonviolence in the Web of Life,* Christopher Key Chapple, ed.
- *Judaism and Ecology: Created World and Revealed Word,* Hava Tirosh-Samuelson, ed.

Ecology and the Environment: Perspectives from the Humanities

Edited by Donald K. Swearer

with Susan Lloyd McGarry

Center for the Study of World Religions
Harvard Divinity School
Cambridge, Massachusetts
Distributed by Harvard University Press
2009

Printed in the United States of America

Grateful acknowledgement is made for permission to use the following:

Donald Worster, "Nature, Liberty, and Equality" appeared in a slightly different form in *American Wilderness: A New History*, ed. Michael Lewis, 263-72. Copyright © 2007 by Oxford University Press. Used by permission of Oxford University Press and the author.

Page 60: "Sitting at Night at Pi-Hsia Pond" by Wang Yangming as translated by Julia Ching in Julia Ching, *To Acquire Wisdom: The Way of Wang Yang-Ming* (New York: Columbia University Press, 1976). Used by permission of John Ching.

Page 113: "Green Turtle" by Michael Jackson, first printed in *Antipodes* by Michael Jackson, 9. Copyright © 1986 by Auckland University Press. Used by permission of Auckland University Press and the author.

Cover photograph: © John Kelly/Getty Images

Cover design: Kristie Welsh

Library of Congress Cataloging-in-Publication Data
Ecology and the environment : perspectives from the humanities / edited by Donald K. Swearer with Susan Lloyd McGarry.
p. cm. -- (Religions of the world and ecology)
Includes bibliographical references and index.
Summary: "Examines ethical, religious, and aesthetic dimensions of the environment from several different disciplines related to the humanities including anthropology, literature, philosophy, religious studies, and history, with examples drawn from Confucianism, aboriginal Australia, Moby-Dick, liberal democracies, Ken Wilber, Joanna Macy, and Gary Snyder"--Provided by publisher.
ISBN-13: 978-0-945454-43-4 (soft cover : alk. paper)
ISBN-10: 0-945454-43-0 (soft cover : alk. paper) 1. Environmental sciences. 2. Environmental ethics. 3. Environmental sciences--Religious aspects. 4. Environment (Aesthetics) 5. Human ecology--Religious aspects. 6. Humanities. I. Swearer, Donald K., 1934- II. McGarry, Susan Lloyd. III. Title. IV. Series.

GE105.E36 2008
304.2--dc22

2008043682

Contents

Foreword

Daniel P. Schrag

It is an interesting time to be an environmental scientist. Fossil-fuel use continues to spew greenhouse gases into the atmosphere, conducting an experiment on the planet Earth that has never been seen in its more than 35 million years. And the alarming signs of global change, fresh observations from oceanographers, glaciologists, ecologists, or meteorologists, which used to pop up every year or two, now seem to greet me every day with my morning coffee: yesterday more bad news about the rate of retreat of Arctic sea ice—some scientists now estimate that the Arctic will be ice free in the summer within the next decade; today new information about the impact of biofuels on food production and deforestation. What will we learn tomorrow?

In my own work on the history of climate change on the Earth, I have started to think more and more about climate-change solutions—about new energy technologies that will lower carbon emissions and provide energy security and also about engineering strategies to protect Earth ecosystems from collapse if climate-change impacts were faster and greater than we expect. In these new endeavors, I speak regularly with economists, with chemists, with biologists, with policy makers, and with business leaders. It seems critical to foster these conversations and interactions between scholars from different fields in the natural and social sciences and between scholars and the people who have the power to take our ideas and turn them into action.

In all of these activities around climate change and energy technologies for the world, it is easy to forget the humanities. After all, I am trying to develop new technologies that will produce clean electricity for the world at a massive scale. I need to know what these investments will cost and how they will affect the other critical aspects of our economy. I need to know how to get good policies implemented—how to spur businesses and governments to understand the issues and overcome their hesitancy to embark on such an ambitious path. There is little time to think about literature, philosophy, and moral reasoning or to study the treatment of the environment in different cultures, or even in our own histories.

Yet, I know that this tendency to overlook the humanities and retreat to my world filled with science and economics is a grave mistake. Facing our environmental challenges requires new technologies and deep understanding of the scientific basis for our woes. It requires understanding world trade and the economic trade-offs that come with various policy choices. But it also requires understanding the cultural components that led us here, the religious and philosophical traditions that affect how people make choices about their interactions with the natural world, and the social norms that are fostered by music, by art, and by literature.

I was reminded of this point during a recent trip to India. Speaking at a conference in New Delhi to a gathering of the most elite members of the government and industry, I was eager to share with them my views on Indian energy investments and how India should think about confronting the climate-energy challenge. The prime minister, the finance minister, and Sonia Gandhi, the leader of the Congress Party, all spoke in the morning, As these very prominent leaders gave their talks and addressed the questions that followed, I suddenly felt completely unprepared. I could certainly speak about climate change, about the dangers of expanding their coal production and the impacts on air quality, on the melting of glaciers in the Himalayas, and on water security. But I realized that I was in a foreign culture and was about to speak with very little understanding of the cultural traditions around environmental matters. Were these people also shaped by the magical writings of Thoreau and Rachel Carson? What were the stories and beliefs that held significance for them? How would these affect their reaction to my words—how they would see this issue that threatens their prospects for rapid development into a prosperous world power?

It is in this context that I encouraged the modest conference out of which this volume arose. In my role as director of the Harvard University Center for the Environment, I sometimes confront an expectation that my job is to create interdisciplinary engagements, in this case between humanists and scientists. I state most forcefully that this is not the case.

Interdisciplinary activities can be fascinating and inspiring, but in my experience, these attempts are very challenging. The style and procedures of one group are often cryptic to the other. Scientists speak extemporaneously, and assume they will be interrupted, challenged, provoked. Humanists often read their talks, for the precision of their language is at the core of their effort. And the challenges extend far beyond style. Trying to blend such diverse scholarly approaches can often water down the product to the lowest common denominator. Both groups are bending so far over to accommodate the other that the magic of each scholarly tradition may get lost in translation, and both groups go away frustrated. I do look forward to trying to overcome disciplinary boundaries and to working to help our community form bridges across different fields. However, it is just as important to encourage the humanities to speak out from within their many different disciplinary homes without need for explanation or translation. The challenges we face are very great. As *Ecology and the Environment: Perspectives from the Humanities* shows, scholars from history, from philosophy, from religion, and from literature can offer important lessons to guide us through murky waters ahead. And it is important for those of us who remain largely ignorant of their lessons to listen and try to understand.

Daniel P. Schrag
Director, Harvard University Center for the Environment
Harvard University

Preface

Mary Evelyn Tucker

This volume considers the responses of the humanities to the immense and interconnected dimensions of our global environmental crisis. This crisis is born of the irony of unintended consequences and invites us to new forms of creativity.

With the rapid population growth in the twentieth century from 2 billion to over 6 billion combined with the collective human effort to spread modern industrialization, human beings have become a planetary force that is now affecting all life forms on Earth. Modernity, no doubt, has brought great benefits, including improved health and rapid communication among cultures and individuals. Yet the unforeseen consequences of the promise of progress and the allure of modernization are now returning to haunt us. In our blind race to build a prosperous world we have inadvertently undermined the very conditions necessary for a sustainable future.

Natural and social scientists have been documenting this undermining process for many decades with extensive research and numerous publications. Authors of the 2005 Millennium Ecosystem Assessment Report observe that degradation of the ecosystems on which all life depends threatens not only human welfare and survival, but also the well-being and survival of cultures throughout the world.[1] Clearly, those of us in the humanities who are concerned with the study of culture and civilization cannot ignore this challenge of sustainability.

This danger calls us to examine the very nature of who we are as humans and what our role is in relation to the natural world. Are we ourselves a life-threatening species or a life-supporting species? Are we emerging as a species who is losing its claim to the name *Homo sapiens*? Can "wise humans" be the ones who are threatening the survival of other life forms, changing the nature of the climate, drying up rivers and aquifers, destroying top soil and forests at a rapid rate?

If we are concerned about these questions, those of us in the humanities need to join in the conversations of the natural and social scientists regarding the environmental crisis in its global and local manifestations. We surely have something significant to contribute.

Philosophers and theologians, historians and artists have reflected for centuries on this question of our nature as humans in a sustained, if contested, manner. Human interdependence with other humans and with nature has been an important focus of these discussions. Moreover, how civilizations are shaped and flourish are key concerns of many in the humanities. Indeed, these explorations have been at the heart of the humanities since the rise of the universities of the West—Paris, Oxford, Cambridge—and in the centers of learning in the Hindu, Buddhist, Confucian, Jewish, Islamic, and indigenous worlds as well. We need to draw on these reflections from the humanistic traditions of the world's cultures and civilizations.

In doing so, we are creating new grounds for a dialogue of civilizations around issues of a sustainable future, not only for humans but for the broader Earth community as well. In this context a sustainable future will depend on the emergence of a pluralistic, multiform planetary civilization concerned with identifying the shape of mutually enhancing human-Earth relations.

At the same time we need to ask ourselves: are we simply engaged in wishful thinking that scholars in the humanities, disciplines that are the least valued and often marginalized in the contemporary academy, can actually enter into conversation with each other and with our colleagues in the natural and social sciences? Even though significant structures of the academy—the silo mentality of disciplines and the tenure reward system—militate against such dialogue and interdisciplinary cooperation?

Yet new conversations are emerging beyond the forces of disciplinolatry, specialization, and pressure to publish. Our common grounds may indeed transcend our differences. This suggests that we could help to usher in broader understandings of diverse historical perspectives, aesthetic sensibilities, and ethical values that will enhance not only environmental studies as it is being conceived within the academy, but also environmental policy as it is reaching beyond the academy. This is no small task and requires sustained commitment. Yet the fact that environmentalists are inviting humanists into these discussions is a cause for some encouragement.

For we are all realizing—humanists and scientists alike—that the question of who we are as humans is central to the possibility of who we will become as a planetary species.

Science and Policy are Necessary but not Sufficient

In this spirit, leaders from both science and policy fields are beginning to analyze our current planetary situation and reflect on why we have not made more progress in solving environmental issues. The enormous contributions of science over the last 50 years to our understanding of many aspects of environmental problems, both global and local, need to be fully recognized. Without the careful and collaborative research of thousands of scientists around the planet we would be virtually blind to the state of the environment and our effects on it. We would be unaware of such macrophase issues as global warming or species extinction and we would, no doubt, be unaware of a range of issues such as pollution and its effects on health.

However, while countless scientific studies have been published and then translated into policy reports, many experts feel we have not made progress in implementing effective solutions. We are stymied by a range of obstacles from lack of political will to ignorance, denial, and inertia. Scientists are noting that dire facts about environmental problems, as alarming as they may be, have not altered the kinds of human behavior that is rapaciously exploiting nature. Nor have such facts affected human habits of addictive consumption, especially in the richer and now in developing nations. Moreover, policy experts are realizing that legislative or managerial regulation of nature is proving insufficient to the complex environmental challenges at hand. Environmentalists are observing that while science and policy approaches are clearly necessary, they are not sufficient in helping to transform human consciousness and behavior for a sustainable future. These thinkers are suggesting instead that values and ethics, religion and spirituality may be important factors in this transformation. This is being articulated in conferences, in publications, and in policy institutes.

Prominent scientists and policy makers are calling for such broad, new thinking to make a transition to a sustainable future possible. They acknowledge that arguments from "sound science" and computer models that draw on reams of data and statistics do not necessarily move people to action. In this vein, James Gustave Speth, the Dean of Yale's School of Forestry and Environmental Studies, in his book *The Bridge at the End of the World* acknowledges that religion, ethics, and values need to play a larger role in environmental discussions.[2] Similarly, the Harvard biologist, E.O. Wilson, in his book *The Future of Life* notes the potential power of religious beliefs and institutions to mobilize large numbers of people for ecological concerns.[3]

Think tanks such as the Worldwatch Institute in Washington, D.C. are also realizing that statistics and alarming reports are not enough to help initiate the changes for an ecologically sustainable world. Denial and paralysis can set in when the future is presented in endless bleak scenarios. In the final chapter of the *Worldwatch State of the World 2003* report, senior researcher Gary Gardner wrote of the growing role of religions in shaping attitudes and action for a broader commitment to environmental protection and restoration. His essay received significant attention and the larger version of the chapter is published in a separate Worldwatch Paper titled "Invoking the Spirit: Religion and Spirituality in the Quest for a Sustainable World."[4]

In the thirty-year anniversary edition of *Limits to Growth* published in 2004, Dennis Meadows and his colleagues observe that we need new "Tools for the Transition to Sustainability." The authors admit, "In our search for ways to encourage the peaceful restructuring of a system that naturally resists its own transformation we have tried many tools. The obvious ones are—rational analysis, data systems thinking, computer modeling, and the clearest words we can find. Those are tools that anyone trained in science and economics would automatically grasp. Like recycling, they are useful, necessary, and they are not enough."[5] Instead, the authors suggest the need for values and ethics beyond the usual frame of environmental science and policy.

Stanford biologist Paul Ehrlich voiced similar concerns in an address to the Ecological Society of America in August 2004. He observed that, "For the first time in human history, global civilization is threatened with collapse....The world therefore needs an ongoing discussion of key ethical issues related to the human predicament in order to help generate the urgently required response." He acknowledged that the Millennium Ecosystem Assessment has undertaken an important evaluation of the conditions of the world's ecosystems. However, he noted, "There is no parallel effort to examine and air what is known about how human cultures, and especially ethics, change, and what kinds of changes might be instigated to lessen the chances of a catastrophic global collapse." He called for the establishment of a Millennium Assessment of Human Behavior (MAHB) to address these problems.[6]

The Emerging Field of Religion and Ecology

Historians and theologians of the world's religions are beginning to make significant contributions to these discussions as the field of religion and ecology is emerging. It is a field still in its infancy and remains to be shaped in a variety of ways and by plural perspectives. It is worth observing that the other humanities—history, literature, and philosophy—are in many respects further advanced than religious studies in environmentally related research, publications, professional associations, and conferences.

The Harvard conference series on world religions and ecology held from 1996–1998 at the Center for the Study of World Religions (CSWR) might be seen as a beginning of this field in terms of the world's religions, although some studies preceded it in Christianity and several of the other world religions. This international conference series involved over 800 participants and resulted in ten groundbreaking volumes that demonstrate how perspectives and values regarding nature are shaped, in frequently contested ways, by various religions, cultures, and geographies. Many of the participants hoped that this broadened perspective would contribute a more comprehensive and culturally diverse basis for environmental ethics as it is conceived both inside the academy and beyond. This is indeed occurring.

With the growing interest in comparative environmental ethics and global ethics these Harvard volumes are becoming useful resources for ethicists, philosophers, and theologians as well as policy makers. Consequently, several of the Harvard books are being translated into other languages, such as Arabic, Farsi, Urdhu, Indonesian, and Turkish in the Islamic world. In addition, the volumes on Confucianism, Taoism, and Buddhism have been translated into Chinese and the Christianity volume will be translated into Spanish. These books thus have the potential to connect to those interested in environmental policy in particular countries and regions around the world. For example, the government of Iran and the United Nations Environment Programme (UNEP) held two conferences on religion and ecology in Tehran in 2001 and 2005. The Islam volume was signaled for attention by the Minister of the Environment in Tehran who is encouraging its translation into Farsi. Similarly, for a number of years, the Vice Minister for the Environment in China, Pan Yue, has delivered numerous speeches calling for the development of environmental ethics in China based on the traditions of Confucianism, Taoism, and Buddhism. In addition, *Hinduism and Ecology* and *Jainism and Ecology* have been published in India where there are numerous projects already underway, such as tree planting and river restoration, based on Hindu and Jain values.

There is still much to be done within the academy for the newly emerging field of world religions and ecology, which has yet to be more robustly shaped and defined. As the field unfolds the role of scholars will also be developing—documenting this newly emerging alliance, identifying the resources for its further emergence, and providing critical analysis. Throughout this process scholars in the field will be striving to raise thoughtful questions and to pioneer self-reflexive methodologies. There are varied roles here for engaged intellectuals, constructive theologians, and historically and textually based scholars. In addition, the challenge is to create bridges between other scholars interested in the environment from the perspective of the humanities and from the social, natural, and even the applied sciences such as medicine and public health. This is why the Forum on Religion and Ecology organized major interdisciplinary conferences in New York City at the United Nations and the American Museum of Natural History in October 1998, attended by over 1000 people. In addition, we held interdisciplinary conferences on world religions and animals, nature writers and the ecological imagination, religion and climate change, and seminars on cosmology and religion.

In this context, religions and scholars of religions can be seen as necessary—although not sufficient—partners because they need also to be in dialogue with scientists, economists, and policy makers. In creating the web site at Harvard under the Center for the Environment (www.environment.harvard.edu/religion), we highlighted these dialogue partners. In addition, we included broad movements beyond institutional religions, such as deep ecology, ecofeminism, and environmental justice, which are helping to shape the discussion of values for a viable future. The key is to create a tent large enough for all of these to coexist and contribute.

The conference series on world religions and ecology sponsored by Harvard's CSWR was based on an acknowledgement of the dark side of religions as well as recognition of the disjunction of religious traditions and modern environmental problems. The participants acknowledged the historical and cultural divide between texts written in earlier periods for different ends. They worked within a process of retrieval of texts and traditions, critical reevaluation, and reconstruction for present circumstances. They underscored the gap between theory and practice, noting that textual passages celebrating nature do not automatically lead to protection of nature. This suggests an important dialogue that should occur between environmental historians and historians of religions to explore the interaction of intellectual ideas and practices in relation to actual environmental conditions.

Despite these caveats, there is growing recognition that religious and cultural traditions have helped to shape worldviews and ethics regarding nature and our place in it. These traditions, moreover, are far from static entities, but rather are dynamic, contested processes, adapting with different times and circumstances. Indeed, our studies in the humanities are dedicated to exploring how religious and cultural traditions are constantly negotiating the boundaries of change and continuity and of ideas and action.

NOTES

1. The Millenium Ecosystem Report is a several volume, multi-year report. This point is made in several places within the report. For example, see Millenium Ecosystem Assessment, *Ecosystems and Human Well-being: Synthesis* (Washington, D.C.: Island Press, 2005), 46 and 120. Most of the report can be accessed online at http://www.maweb.org/en/Reports.aspx.
2. James Gustave Speth, *The Bridge at the End of the World: Capitalism, the Environment, and Crossing from Crisis to Sustainability* (New Haven, Conn.: Yale University Press, 2008).
3. Edmund O. Wilson, *The Future of Life* (New York: Knopf, 2002).
4. Gary Gardner, "Invoking the Spirit: Religion and Spirituality in the Quest for a Sustainable World," Worldwatch Paper 164 (Washington, D.C.: Worldwatch Institute, 2003).
5. Donella Meadows, Jorgen Randers, Dennis Meadows, *Limits to Growth: The 30-year Update* (White River Junction, Vt.: Chelsea Green, 2004), 269.
6. The remarks quoted come from Paul Ehrlich's address to the eighty-ninth annual meeting of the Ecological Society of America, Portland Ore., August 2004 (text available online at the University of Stanford news service, http://news-service.stanford.edu/news/2004/august4/esa-84.html). He reiterated this call in *Science*. See Paul Ehrlich and Donald Kennedy, "Millennium Assessment of Human Behavior," *Science* 309 (July 22, 2005): 562–63.

Introduction

Donald K. Swearer

The broad popularity of Al Gore's *An Inconvenient Truth* demonstrates the widespread belief—and possibly fear—that environmental change has reached crisis proportions. Indeed, eco-apocalyptic writing has become a popular genre in the burgeoning field of environmental literature; even those who continue to question human causes of global warming cannot escape the reality of record-setting temperatures, violent and erratic storms, melting glaciers, and rising sea levels; and, at colleges and universities demand for more environmental studies courses grows exponentially. Whether in the public sphere or academic context, however, the natural and social sciences have dominated environmental discourse, academic agendas, and government policies. Although there are exceptions, humanists have been at the margins.

The papers assembled in this volume were presented at a conference sponsored by the Center for the Study of World Religions (CSWR) at Harvard Divinity School and the Harvard University Center for the Environment (HUCE) in March 2006. The intention behind this joint venture was to highlight the diverse range of humanistic perspectives, broadly conceived, relevant to debates about the environment, environmental policy, and environmental studies. Roger Kennedy, Director of the National Park Service during the Clinton administration, put it aptly during the conference when he observed that "environmental policy isn't just Gifford Pinchot telling people to do things; it's listening to John Muir's poetic musings about Yosemite, as well."[1] And, during the discussion following the presentations, one conferee commented that the humanities might rescue the policy and scientific communities from the ideology of "techno-scientific-salvationism," the belief that science will solve the environmental crisis. Such a belief was expressed in the Summer 1996 issue of *Daedalus* on science, technology, and the environment which—recalling the epigraph inscribed on the dome of the National Academy of Sciences building in Washington, D.C. ("To science, pilot of industry, conqueror of disease, multiplier of the harvest, explorer of the universe, revealer of nature's laws, eternal guide to truth")—opined, "We have liberated ourselves from the environment. Now it is time to liberate the environment itself."[2]

The conference presenters were invited to speak from their particular areas of expertise and disciplinary perspectives within the humanities and cultural studies: literature, history, religion, philosophy, ethics, and anthropology. The papers vary in several respects, most importantly the range of perspectives they bring not only to the scope of environmental studies but also in regard to the questions they raise about the direction of environmental policy. Obviously, the humanities are not monolithic, as the conference papers so well demonstrate. Although the humanities do not speak with a unified voice, they broaden and deepen our understanding of the natural world and the relationship of the human community to it. How we narrate, mythologize, and philosophize about the environment along with the religious and ethical values that we ascribe to the natural world are not only frames of imaging and understanding but of acting and living in the world. The essays in this volume offer a kaleidoscopic view of what Michael Zimmerman in this volume characterizes as an "integral ecology." I shall briefly comment on each of the essays.

A leading interpreter of American nature writing, Lawrence Buell, Harvard University, points out that the question of art's practical use-value has always been problematic. He argues for the necessity of "environmental imagination,"[3] or as he puts it in his opening conference remarks, "the arts of imagination," in appreciating and valuing the environment. For example, a land developer considers a wetlands as a construction site that needs to be drained in order to be useful, while a poet brings a sense of environmental aesthetics that envisions the wetlands as part of a larger ecosystem, a palace of biodiversity, and an inspiration for eco-narratives framed in images and metaphors rather than the monetary calculus of profit and loss.[4] In support, Buell quotes Ulrich Beck, "Only if nature is brought into people's everyday images, into the stories they tell, can its beauty and its suffering be seen and focused on." Buell briefly explores

Melville's *Moby-Dick* to illustrate how art and literature might script environmental-ethical concern as an exploratory thought experiment rather than as a political agenda. He cautions against an excessively utilitarian or political view of the artistic and literary contribution to environmental studies and environmental policy: the demand to impose a yardstick of manifest advocacy in service to a cause, rather than as a product of polymorphous intellectual curiosity and a broad moral concern for the environment.[5] The moral sentiment that undergirds the environmental imagination cannot be quantified; rather, it informs environmental discourse from a deeply felt and broadly nuanced aesthetic-moral sensibility.

From a historical perspective, Donald Worster, Department of History, University of Kansas, sees a strong connection between the traditions of liberal democracy and nature conservation, not simply from a political or policy perspective but from deeply held humanistic principles and values. Worster takes issue with the argument that only rich countries can afford conservation, noting that a country like Costa Rica ranks 90th in per capita income but has protected 28 percent of its territory from development. He argues rather that nature protection occurs within nations that profess democratic principles, cherish human rights, and uphold the freedom of speech. Worster cites the U.S. as noteworthy for its long history of environmental protection, a tradition under threat today from the political right. Environmental protection has failed when confronted by authoritarian regimes such as Somoza in Nicaragua and Pol Pot in Cambodia: as Worster succinctly observes, "Nature is a threat to the authoritarian mind." In support of this view, Mary Evelyn Tucker observes that the greatest environmental destruction in China occurred under Maoist rule.

Tucker is one of the leading American academics promoting the field of religion and ecology. She and John Grimm organized eleven conferences with the CSWR between 1996 and 1998 that led to the founding of the Forum on Religion and Ecology.[6] A student of Thomas Berry, Tucker has been a strong advocate of the contribution that religious worldviews and practices can make to influencing human attitudes and actions regarding the natural environment based on awareness of the essential interconnection of all life forms.[7] Her preface places the field of religion and ecology within the current interdisciplinary dialogue aimed at creating a sustainable future for the broader earth community. Tucker's paper in the volume delineates the relationship between nature and Neo-Confucian ideology, her area of scholarly research. She argues that the three pillars of the Neo-Confucian philosopher, Wang Yangming (1472–1529)—empathetic knowing, embodied acting, and compassionate living in the world—situate a social ethic within an ecological philosophy that stresses the reciprocal nature of all life forms. Empathetic knowing affirms the subjective and qualitative as the primary way of apprehending the nature of things rather than quantitative measures; embodied acting integrates holistic understanding and action based on that knowledge; and compassionate living embraces a common kinship with a larger community of life beyond *Homo sapiens*. Wang's life and teaching exemplify a sympathetic resonance with all things, an "interpenetration of self, society, nature, and cosmos." In Wang, Tucker finds lessons for an environmental ethic based on a naturalistic cosmology, a worldview of organic holism, a vision of the continuity of being, and an internal process of cultivation that has relevance for engaging the world. In Wang's China, these were not simply abstract ideals, she contends, but were promoted practically through institutions and humanistic endeavors: academies, universities, libraries, printing, and the arts.

In related yet significantly different ways Bron Taylor and Michael Zimmerman also stress the importance of a holistic, integrated, ecocentric worldview for the formation of an environmental ethic. Michael Zimmerman, director of the Center for the Humanities and the Arts at the University of Colorado and former codirector of Tulane University's Environmental Studies Program, approaches environmental ethics from the perspective of Western philosophy. As background to his constructive proposal of "integral ecology," he briefly surveys philosophical issues that anticipate integral ecology such as challenges to the fact/value, moral-ought/prudential-ought distinctions; the Kantian a priori transcendent universal; the romantics' reaction against dualism; and Schelling's identity of nature and spirit. Zimmerman proposes an ecological worldview that ascribes a depth dimension to the cosmos that reflects current research into the sentience of animals and even plants. Zimmerman's constructive project is indebted to the work of Ken Wilber that integrates mind and body, inner and outer, with a particularly strong emphasis on interiority that resonates with Wang Yangming's empathetic knowing: a first-person or interior, intersubjective perspective in contrast to a third-person, objectifying, instrumentalist perspective identified with the natural and social sciences. Integral ecologists bring into an environmental ethic what has often been omitted, namely, the interior, subjective domain that they believe characterizes all phenomena. An environmental ethic based on such an understanding is holistic: inclusive, multidimensional, and multidisciplinary. In Aldo Leopold's *A Sand County Almanac* (1949), Zimmerman finds concrete expression of the first two ways of knowing in Wilber's tripartite schema—(1) self, subjectivity, aesthetics, sincerity; (2) ethics, culture, worldview, intersubjective meaning; (3) natural and social sciences, propositional truth—because Leopold's land ethic takes into account not only material and economic aspects of the land but also aesthetic and ethical aspects. Leopold valued the land for its beauty, integrity, and stability; and, moreover, came to an unusual appreciation of the interiority of nonhuman animals.

Bron Taylor, founder of the new PhD program in Religion, Culture, and the Environment at the University of Florida, also invokes Leopold. His paper schematizes a wide range of environmental thinkers, movements, and activists ranging from Darwin to the eco-spirituality of Joanna Macy in terms of four descriptive types of "dark green" religion ("religion that considers nature to be sacred, imbued with intrinsic value, and worthy of reverent care"): Spiritual and Naturalistic Animism; and Gaian Spirituality and Naturalism. Fluid and overlapping, these categories function as affinity groupings in terms of which Taylor analyzes several movements and actors representing "religious environmentalism:" Gary Snyder, Joanna Macy, John Seed, Marc Bekoff, Jane Goodall, L. Freeman House, Tom Regan, Aldo Leopold, and James Lovelock. Taylor offers these case studies as evidence of the increasing turn to environmentalism by religionists and ethicists and the influence they will have in addressing environmental agendas in the future. He predicts that "dark green religion," a form of religious environmentalism largely "untethered" from organized religion, will play an increasingly important role in global environmental politics. Despite a flavor of utopianism, Taylor's prognosis might be seen as a radical expression of Worster's historically grounded argument for a conservation ethic embedded in the democratic ideals of liberty and equality. Although Taylor's prediction regarding dark green religious environmentalism may come true, today the political role of dark green religious movements and actors is less evident than a "tethered" expression of religious environmentalism, such as the environmental movement within contemporary evangelical Protestantism that has become a potent influence in American politics.

In contrast to the emphasis on holistic, ecocentric worldviews in the papers by Tucker, Zimmerman, and Taylor, Michael Jackson, a social/cultural anthropologist at Harvard Divinity School offers a place-based, ethnographically rich description of the intimate relationship between nature and culture among the Kuku-Yalanji in southeast Cape York, Australia. Among the Kuku-Yalanji, the land is seen as a social reality and nature is viewed through the lens of cultural categories. Nature and culture are imbued equally with a life force that intimately connects human actions and the actions of nature. While seemingly vastly different from Wilber's tripartite schema, Jackson's ethnographic vignette illustrates the first (subjective, self, aesthetic) and second (intersubjective meaning, culture, ethics, worldview) dimensions discussed in Zimmerman's paper: for Aboriginal people the land is steeped in memories of births, deaths and marriages, seasonal movements, and traumatic disruptions; it is a *vita activa*, a process of living and moving with others on the land and drawing one's livelihood from it that charges the landscape with vitality and presence. Jackson's essay challenges philosophers, ethicists, and historians of religion to contextualize the language of worldview in history, culture, and concrete, practical, real-life situations. The ethnographer, in turn, is challenged to transpose culturally thick descriptions into categories that can engage cross-cultural, interdisciplinary environmental discourse.

To illustrate the interconnection between epistemology, ethics, and concrete, practical, real-life situations, I turn to two contrasting examples from my own research in Thai Buddhism. The first is the case of a 1986 project to build a cable car to the top of Mount Suthep which overlooks Chiang Mai, the major metropolis in northern Thailand; the second is the economic theory known as "sufficiency economy" that emerged in Thailand in response to the 1997 Asian financial crisis and is associated with Thailand's reigning monarch, King Bumiphol Adulyadej.

Mount Suthep (*Doi Suthep*) rises 5,250 feet above the city of Chiang Mai, Thailand's third most populous city and since the late thirteenth century the political and cultural center of northern Thailand.[8] Physically the mountain orients the valley's inhabitants; ecologically its watershed sustains an ever growing population; and its forest cover is home to an impressive diversity of flora and fauna that includes over 253 species of orchids, 320 bird and 50 mammal species, and more than 500 species of butterflies. Species of previously unknown plants and animals continue to be discovered on Mount Suthep. The mountain is the heart of a national park covering approximately 100 square miles.

The mountain region has a rich mythological past, which features Wat Phra That Doi Suthep, one of the most revered Buddhist sanctuaries in mainland Southeast Asia, situated near its summit. Myths and legends tell of the authochthonous guardians of the mountain, Phu Sae and Ya Sae; the Brahmanical hermit, Vasudeva, after whom the mountain is named; and conflict between Vilangkha, the chief of the native Lawa, and Queen Cama, who ruled Haripunjaya, the first kingdom in the Chiang Mai valley established in the eighth century. Of surpassing historical and cultural significance, however, is Wat Phra That. Here myth and legend conjoin to become history. Tradition says that the sanctuary was established in the fourteenth century to house a Buddha relic brought by the monk, Sumana Thera, from the Thai kingdom of Sukhothai to Chiang Mai at the request of its ruler, Phaya Kuena (reign dates: 1355–1385 CE), a great patron of the Buddhist religion. These mythic, legendary, and historical narratives provide a cultural map that overlays Mount Suthep's imposing physical presence and around which the history of northern Thailand unfolds. Both physically and symbolically Mount Suthep and other nearby peaks are prominent in the northern Thai cultural imagination and sense of identity. However, have these mountains, once so rich with meaning, lost their power and significance, and become mere tourist landscapes and spaces to be exploited commercially?

The contemporary significance of Mount Suthep as a sacred mountain, natural habitat, and a work of culture became abundantly clear in 1986 during a controversy over the construction of an electric cable car from the base of the mountain to the monastery-temple, Wat Phra That, near the summit. The cable car, endorsed by the Tourist Authority of Thailand, would accommodate an ever-increasing number of tourists who flock to Thailand's northern mountains. Long gone are the days when pilgrimage to Wat Phra That was on foot, but the paved two lane road to the sanctuary built in 1934 with voluntary manual labor under the inspired leadership of the charismatic Buddhist monk, Khruba Siwichai, has itself become part of the mountain's legendary history. However, the proposed cable car to be constructed by a commercial company to exploit Doi Suthep and promote the increasingly invasive commercial degradation of the mountain was another matter. Environmentalists, university professors, students, and ordinary citizens rose up in protest. A key element in quashing the plan was the role played by Buddhist monks, notably the late Phothirangsi, then assistant ecclesiastical governor of the province of Chiang Mai. Sentiments in defense of Mount Suthep as a sacred place in the face of the onslaught of commercial development and tourism identified the mountain iconically with Buddhism. The following paragraph from Niranam Khorabhatham's editorial in the April 30, 1986, *Bangkok Post*, illustrates the tenor of the rhetoric and the deep reverence for the mountain:

> The manager of the proposed cable car project on Doi Suthep, Chiang Mai, states that he was 'not overlooking the sanctity of Wat Phra That.' He underestimates the northern people: The Soul of Lanna [northern Thailand] is still alive. Northerners perceive, at least in their sub-

> conscious, that Mount Suthep is like a symbolic stupa. Doi Suthep's dome-like shape is like an immense replica of the ancient Sanchi style stupa, a gift to Lanna by the Powers of Creation. Stupas are reliquaries of saints [i.e. the Buddha]. More than that, they are a structural representation of the very essence of Buddhism. Plant and animal life are like Nature's frescoes, both beautifying and exemplifying the Law [dharma] not less than paintings in any man-made shrine. Although sometimes not being able to explain why rationally, the northern people want to preserve the Stupa Doi Suthep as it was given to them by Creation, as untouched as possible, as sacred.

The case of Mount Suthep and the proposed cable car to its summit resonates with Lawrence Buell's plea for "environmental imagination," although in this example the moral sentiment undergirding the imaginative, iconic depiction of the mountain led to direct action. It also serves as a concrete expression of Michael Zimmerman's philosophy of the interior and intersubjective dimensions of integral ecology and of the intertwining of nature and culture in Michael Jackson's study of the Kuku-Yalanji of southeast Cape York, Australia. The pressures to exploit Mount Suthep for its tourist value threatened the mountain's natural environment and its cultural and religious integrity. The fact that Mount Suthep is perceived by northern Thais as a sacred landscape was a major factor in challenging both private and government plans to build a cable car to the top of the mountain, and to expand tourism and other commercial enterprises destructive to its natural environment. Reverence for Mount Suthep in the cultural and environmental imagination of northern Thais and its association with Buddhism served as the basis for a conservationist environmental ethics.

For my second example I turn to the philosophy of "sufficiency economy," initially articulated by King Bhumipol in speeches beginning in 1974, and its broader relevance to planetary sustainability.[9] Prompted by the Asian financial crisis of 1997, the National Economic and Social Development Board of Thailand invited several of the country's leading social thinkers to develop more systematically the king's proposals regarding a more balanced, sustainable development in response to globalization. Since the 1950s the government had prioritized development; in the 1970s foreign investment from the United States and other countries burgeoned; by the early 1990s a new Japanese factory opened in Thailand every three days, and approximately a million people left agricultural production for urban jobs every year. This top-down development model led to impressive Gross Domestic Product (GDP) growth that averaged 7.6 percent from 1957 to 1997. However, the capital, technology, and techniques came from the outside; rural debt mushroomed; and the risk and fluctuation of world markets on which Thailand's economy depended increased material vulnerability, mental anxiety, and social anomie as economic and social forces careened out of control. Furthermore, the pace and rapacity of growth resulted in severe environmental degradation. Between 1947 and 2000 two thirds of Thailand's forests disappeared contributing to a significant increase in flooding and landslides. Furthermore, industralization and urban growth has led to environmental problems of waste disposal, pollution, and conflict over water resources.

Beginning as early as the 1960s a "discourse of discontent" arose out of concern over the destructive, divisive, unsustainable, and disempowering byproducts of growth. This discourse emphasized the need to rebuild a sense of community which had been undermined by the global marketplace, to develop greater economic self-reliance, to restore horizontal networks of knowledge and distribution, and to draw on the teachings of Buddhism with its emphasis on moderation and spiritual well-being as an antidote to maximizing growth and consumption. The shift from consumption to "people-centered development" emphasized "education, healthcare and social welfare; equitable sharing through regionalization, participation and community rights; and rehabilitation of environment through better management and greater local control."[10]

The philosophy of sufficiency economy reflects Buddhist ideals framed as general principles. The 2007 UNDP Thailand Human Development Report stipulates these principles as follows:

- everyone encounters suffering but every individual has the mental ability to eventually rise above it;
- maximizing consumption does not lead to happiness beyond a certain point, wastes finite resources, and engenders competition that leads to conflict;
- an economy works more harmoniously based not on selfish greed but principles of empathy, compassion, fairness and generosity, honesty, straightforwardness, and a refusal to exploit others; and
- Buddhist economics is based on the principle of the middle way which balances the interest of self and others; a theory of rationality that acknowledges the interrelationship between means and end, motive and result; and the recognition of the constitutive interdependence of all life forms.[11]

The philosophy of sufficiency economy is not a theory of Buddhist economics per se, but advocates for a holistic, integral, view of what it means to be human in the broadest sense of biotic community.

The Harvard conference aimed to give voice to the humanities as a full partner with the natural and social sciences in comprehending the complexity and multidimensionality of the global environmental crisis and its possible solutions. The insights that might result from broad interdisciplinary environmental dialogue emerged during the discussion following the conference presentations. The relevant cogency of integral, ecocentric worldviews and communities of interpretation illustrated by the conference papers was addressed by a biological ecologist in the audience who observed that *umwelt* worldviews are the stuff of being human measured biologically in the genes' cytoplasm that make up human beings. That the current generation has 250 chemicals in its cytoplasm, particularly adipose tissue our grandparents did not have, bears witness to the fact that being a human being or a species involves an exchange with environmental, physical, and chemical factors. Furthermore, genetic shifts in human beings are shared with other species, not just those closely related to us but house mice and others, as well. Another example is the exchange of epizootics such as the avian flu, illustrating not only the interaction of another genome transported across the globalized world, but the open-window, interrelational nature of what it means to be human. As this discussion demonstrates, current biological views of the ways in which our genome and cytoplasm reflect the environment can be nuanced, enriched, and challenged by holistic, ecocentric worldviews, cultural narratives, and literary "imaginings" over and against those forms of environmental discourse predicated on instrumentalist approaches to the environment, ecological communities, and human nature.

The conference would not have taken place without the support and input of its codirectors, Professors Daniel Schrag, an expert on earth planetary science and climate change, and Lawrence Buell, the Powell M. Cabot Professor of American Literature at Harvard. As director of the Harvard University Center for the Environment, Professsor Schrag has been a strong proponent for increasing the presence of the humanities in Harvard's environmental studies program; and, from the early planning stages, Professor Buell's broad interdisciplinary knowledge of the fields of American nature writing and environmental studies was indispensable to the direction and shape of the conference. To both of them my heartfelt thanks.

I also wish to express my deepest appreciation to Susan Lloyd McGarry, the CSWR Manager of Planning and Special Projects, without whose editorial assistance and oversight this volume would not have been possible.

Donald K. Swearer
Director, Center for the Study of World Religions
Harvard Divinity School

NOTES

1. Comments by Kennedy in the uncorrected unpublished transcription of the conference, March 2006. In 1905 Pinchot was appointed the first chief of the United States Forest Service.
2. Jesse H. Ausubel, "The Liberation of the Environment," *Daedalus* 125, no. 3 (Summer 1996): 15.
3. Lawrence Buell, *The Environmental Imagination: Thoreau, Nature Writing, and the Formation of American Culture* (Cambridge: Harvard University Press, 1995).
4. See John McPhee, *Encounters With the Archdruid* (New York: Farrar, Straus and Giroux, 1977).
5. See Buell's essay in this volume.
6. See the FORE website (www.religionandecology.org).
7. Thomas Berry, *The Dream of the Earth* (San Francisco: Sierra Club Books, 1988); Mary Evelyn Tucker, *Worldly Wonder: Religions Enter Their Ecological Phase* (Chicago & LaSalle: Open Court, 2003).
8. The following discussion of *Doi Suthep* and the 1986 cable car case is adapted from Donald K. Swearer, Sommai Premchit, and Phaitun Dokbuakaew, *Sacred Mountains in Northern Thailand and Their Legends* (Chiang Mai, Thailand: Silkworm Books, 2004) and Donald K. Swearer, "Principles and Poetry, Places and Stories: The Resources of Buddhist Ecology," *Daedalus*, 30, no. 4 (Fall, 2001): 225–241.
9. The following discussion is taken from the *UNDP Thailand Human Development Report 2007: Sufficiency Economy and Human Development* (Bangkok: United Nations Development Programme, 2007), chapt. 2, "Thinking out the Sufficiency Economy," 20–35.
10. Ibid., 25.
11. Ibid., 31.

Literature as Environmental(ist) Thought Experiment

Lawrence Buell

"What, if anything, do literature and the other arts, and scholars of them, have to bring to the table at a time of environmental crisis?" This is a question I am often asked, within the academy as well as outside—and understandably so. The public at large looks to the arts and humanities, when it looks at all, for recreation or off-duty cultural "uplift" rather than for "solutions"; and within the academy, those who study, teach, and write about the environment are predominantly scientists, engineers, economists, lawyers, medical researchers, or public policy specialists, for whom the methods of humanistic research tend to seem mysterious if not positively suspect, and the sphere of the arts tangential if not irrelevant to their own central tasks as professional environmental researchers. Environmental humanists are still by contrast few, though our ranks are growing.

To such skeptical questions, there is an easy and, up to a point, cogent first-stage retort: namely, that the arts of imagination, and the formal study thereof, have a far more important role to play than the questioner's usually needling tone implies. How do you turn a "swamp" into a "wetland"? The transition from the bygone era of promiscuous swamp-draining to the contemporary age of comparative wetlands protection has not happened as a result of scientific expertise alone or through litigation alone. It has also required a broader—still ongoing—transformation of public values and commitments, to which end the arts of language and imaging have been indispensable, starting with the question of the signifier of preference (benign "wetland" versus pejorative "swamp").[1] Of all those strongly opposed to Alaska North Slope drilling, how many are moved by media images of northern Alaska as the nation's last truly pristine large tract of wilderness? Likely far more than those who have actually been there or who have studied the arguments pro and con with care and/or expertise. The Rachel Carson of contemporary social theory, Ulrich Beck, rightly declares that "Only if nature is brought into people's everyday images, into the stories they tell, can its beauty and its suffering be seen and focused on."[2]

Small wonder, then, that advocates for the sport-hunting industry used to complain that the worst thing that had ever happened to it was Bambi.[3] Small wonder that Carson's *Silent Spring* is agreed to have had a much more significant role in galvanizing 1960s' activism around pesticides and environmental pollution rather than the two other contemporaneous, also-hard-hitting books on the same issue: Murray Bookchin's *Our Synthetic Environment* and Robert Rudd's *Pesticides and the Living Landscape* (sometimes called "*Silent Spring* for professors.") The difference lay not just in Carson's better networking into the publishing industry (serialization in *The New Yorker*, for example), but in her far better eye for the telling story and the unforgettable image: in short, in her literary edge as an award-winning nature writer.[4]

So there is a strong *prima facie* case for the value-changing potential of environmental aesthetics. But the case as just framed is also incomplete and question-begging. Grant the existence of demonstrable causal links between certain acts of environmental imagination and individual life-practice, environmental reform movements, and reform legislation. Grant Henry Thoreau's inspiration for modern back-to-the-landers and preservation efforts in Boston's backyard, as well as elsewhere in New England and beyond. Grant the impact on Progressive-Era reform legislation of Upton Sinclair's novelistic exposé of the Chicago meatpacking industry, *The Jungle* (1906). Grant the significance of Carson's *Silent Spring* (1962) in effecting the ban on DDT and early 1970s environmental protection legislation more generally. Grant the influence of Edward Abbey's passionate, rompishly outrageous novel *The Monkey Wrench Gang* (1975) about eco-sabotage "as a prototype for the development of Earth First!"[5] Grant that future acts of imagination in text, film, and other media will now and again have similarly great or even greater impact. However true and important this may be, however exciting the prospect, if that is the one thing we can say about the significance of environmental art and criticism, we risk reducing art to instrumentalism and scholars to cheerleading chroniclers of a limited number of exceptional cases.

This issue of art's use-value has always been hard for either artists or critics to discuss without falling into hypocrisy or self-contradiction. "How many a man has dated a new era in his life from the reading of a book," declares Henry Thoreau in *Walden*—clearly hoping his book will have such an impact. On the other hand, elsewhere he no less categorically insists, "If I knew for a certainty that a man was coming to my house with the conscious design of doing me good, I should run for my life."[6] Surely someone capable of thinking like that would, in some compartment of his mind at least, have felt ambivalent about seeing his masterpiece turned into a template or poster child for an environmentalist or back-to-nature program of whatever sort, however worthy—ignoring his penchant for sententiousness—forever seeming to lay down the law in catchy, prescriptive aphorisms.

As with creative writers, so with critics, and never more so than now, when there is so strong a push within academia across the board to want to coordinate, if not positively conjoin the roles of scholarly investigator and public intellectual-activist. In my own particular bailiwick of environmental humanities, literature and environment studies or "ecocriticism," two salient, often intertwined, ways in which this border-crossing is being urged and done right now, are so-called "narrative scholarship," (i.e., the coordination of critical reflection with autobiographical narrative so as to give what otherwise might seem detached scholarly analysis greater immediacy by underscoring the sense of personal witness that further dramatizes claims of eco-stewardship); and environmental justice revisionism, from which standpoint scholarship and direct experiences of advocacy on behalf of immiserated communities ought to go hand in hand. Broadly speaking, the first of these initiatives would push scholarship more toward a celebratory poetics of nature, the latter more toward oppositional social critique; but both rely to such an extent on appeal to the force of stories of personal witness that it would be more accurate to think of the second as a "second wave" reaction from and against the first.[7]

Both have produced, at best, valuable and enlivening results. They can also take a good thing too far. It ought to be possible to credit art and scholarship with being seriously engaged with issues of social and planetary import short of imposing a yardstick of manifest advocacy. For one thing, worthily eloquent voices may be ignored. For another, influence is notoriously hard to pin down. And even when it can be, it may differ drastically from what was intended. Sinclair's *The Jungle* is a striking case in point. The novel drew the immediate attention of President Theodore Roosevelt, who dispatched a blue-ribbon commission to Chicago, as a result of which, within six months of the book's publication, both the Pure Food and Drug Act and the Beef Inspection Act were passed. Not bad, one might say. This was not, however, the reform the author hoped for. "I aimed at the public's heart, and by accident I hit it in the stomach," he complained. His aim—clear enough from the book itself—was to write "the *Uncle Tom's Cabin* of wage slavery."[8] But the upshot was not to rouse sympathy for the stockyard workers' plight so much as consumerist fears of tainted food. When Sinclair in old age was invited to the White House by Lyndon Johnson to witness the signing of a new, stiffer meat inspection law, his sense of irony must have been great.[9]

But the prior and more basic problem with impact-oriented thinking is the presumption that environmental writing and criticism ought to be conceived more in terms of service to cause than as the upshot of polymorphous intellectual curiosity or diffuse concern for environmentality not associated with any settled position. At the birth of the ecocritical insurgency in the early 1990s, a strong activist and normative impetus was doubtless needed in order to make a forceful debut, as with first-wave race and feminist and sexuality studies; but programmatic initiatives of whatever stamp all too easily become more a constraint than an enabler.

In a cross-disciplinary forum on religion, values, and the environment this point deserves special emphasis because the residual temptation to second the chorus of well-meaning nonspecialists and laypersons (such as Ulrich Beck) who tie art's value to its social use-value is easily reinforced by the temptation in the present context to extract from works of art or map them onto paradigms of environmental value. In such a venue, one cannot help but feel sheepishly untoward or downright uncollegial by declining to hold up Thoreau, for example, as a prophet of biophilia and biodiversity, as E. O. Wilson does in *The Future of Life*;[10] or Thomas Hardy's Wessex novels as harbingers of bioregionalism; or Upton Sinclair as a forerunner of the environmental justice movement. All of these diagnoses can, indeed, be made to stick. Yet, it is the suggestiveness and intricacy with which such dispositions are engaged or adjudicated rather than any decisive commitment to them that is likely to keep an act of environmental imagination alive. Likewise, ecocriticism's own staying power and percolation effect, certainly within the academy and also in the long run beyond it, will most likely depend less on its categorization of a text's environmental(ist) ethics or politics, or on how it does or does not fit the critic's own conception of same, than on its ability to demonstrate the hitherto-underperceived significance of environmentality as a pervasive, fascinating, compelling concern in world literature from the beginnings.

And this will only happen when more of us do a better job of talking about how art and literature script environmental-ethical concern, not as if it could be translated readily into the terms of prevailing ethical or political programs but as a thought experiment: i.e., in exploratory, often tentative ways complicated by multiple agendas and refusal to take fixed positions.

Elsewhere I have discussed more pointedly my own preferred method for parsing literature in this way,[11] which involves situating texts within a conceptual framework that draws eclectically on phenomenology of perception, intellectual history, humanistic geography, science studies, genre theory, and cultural/ideological criticism. The payoff always lies in how any such apparatus can help disclose the intimations of environmentality in individual texts and their horizons of perception and implication: how they adumbrate, refract, and engage environmental values of whatever sort—more often tentatively or playfully rather than with a true-believing commitedness; and, like any act of human reflection, with something less than 360-degree awareness of all that is potentially at stake: an awareness that is "prophetic" in the sense of being suffused with visionary energy rather than in the conventional sense of precise diagnosis or precision.

In order to instantiate this perspective, I turn to a passage from a widely familiar text, Herman Melville's *Moby-Dick* (1851).[12]

This passage takes off from the question of why—or so the speaker claims—whale meat seems unappetizing to most palates, even whalemen's, hypothesizing that the key reason is not the obvious one—namely, that it is unctious and fatty—but a sense of taboo: "that a man should eat a newly murdered thing of the sea, and eat it too by its own light." But, he continues, in a strange mixture of vehemence and offhandedness:

> . . . No doubt the first man that ever murdered an ox was regarded as a murderer; perhaps he was hung; and if he had been put on his trial by oxen, he certainly would have been; and he certainly deserved it if any murderer does. Go to the meat-market of a Saturday night and see the crowds of live bipeds staring up at the long rows of dead quadrupeds. Does not that sight take a tooth out of the cannibal's jaw? Cannibals? Who is not a cannibal? I tell you that it will be more tolerable for the Fejee that salted down a lean missionary in his cellar against a coming famine; it will be more tolerable for that provident Fejee, I say, in the day of judgment, than for thee, civilized and enlightened gourmand, who nailest geese

> to the ground and feastest on their bloated livers in thy pate-de-foie-gras.[13]

This passage seemingly aims to unsettle standard thinking about two conventional but arbitrary borderlines by questioning each in turn and by crunching the two queries together. It is both a pre-Darwinian interrogation of species borderlines (are humans as different from whales and other nonhumans as they like to think?) and a pre-Boasian questioning of what, if anything, differentiates so-called civilization from so-called savagery. The speaker insinuates that carnivirousness among "civilized" cultures which batten on beef and goose liver is cannibalism in denial, and as such, is arguably worse than the practices that so-called civilizations stigmatize as taboo.

What is much less clear is whether a serious ethical position is being staked out and defended. The passage is a riff on orthodox Christian doublethink that hovers between preachment and Rabelesian grotesquerie. The question of how seriously to take the assault on borderlines is further beclouded by the fact that the text speaks through the mask of a stand-in, the mentally hyperactive serio-comic narrator Ishmael, who is not equatable one-to-one with the author. On the other hand, sorties of this kind, once taken and reiterated are bound to get picked up sooner or later by those who listen. Hence cetacean biologist Roger Payne's praise for Melville's uncanny sense of "just how and by what steps whales would enter our minds, and how once inside they would metastasize . . . throughout the whole engine of human ingenuity, mastering and predisposing it to their purpose."[14] What Payne seems to have in mind here are the many moments when the novel steps aside from its search-and-get-destroyed plot to muse across the species borderline about cetacean intelligence and cetacean anatomy in relation to the human so as to produce a long series of ad hoc exercises in comparative ethology that make it impossible not to think about whales as fellow creatures, rather than bestial or symbolic adversaries: musings on how it might feel to lack binocular vision; to breathe only one seventh of your existence; to have the sense of touch concentrated in a single organ, the tail; to exist in a body that seems skinless from a human perspective yet adaptable to extremes of heat and cold, and so on.

By no means will every reader want to read this novel in this way. Precisely because passages like those just noted are asides, it is quite possible to slide by them. This is the way it has been for almost a century and a half since *Moby-Dick*'s publication: both the nearly seventy-five years during which the novel fell into obscurity and the sesquicentennium since, during which as one waggish fellow critic writes, *Moby-Dick* hermeneutics has superseded whaling as one of New England's leading industries. Most card-carrying Melvillians have preferred to read the novel as a revenge tragedy, as an epistemological meditation or another sort of religiophilosophical inquiry, or (currently most fashionable) as a political or economic allegory of some kind. This, of course, is where environmental criticism comes in: by teasing out the landmark significance of the fact that *Moby-Dick* is the first classic in Anglophone literature to center on a nonhuman creature, to insinuate again and again the porousness of the species borderline between cetacean and human, and to speculate about its ethical implications.[15]

This line of reflection stands to make a permanent enhancement or redirection of readerly thinking only on condition that it conceives of the novel's incipient biocentrism, such as it is, as being in the nature of a thought experiment across species lines, a question-raising about the shortsightedness of speciesism (and, ethnocentrism too), rather than as unfolding a crystallized or consistent environmental ethics; and, furthermore, that this line of reflection is neither the novel's only nor most ostensibly central aspect. It ought to be considered breakthrough enough to change the conversation to the point that it becomes harder and harder not to notice the environmental thought experiment in play. This I take as where the epicenter of ecocritical work ought now to be: the rereading of the corpus of world literature from Gilgamesh to Beowulf on down, as raising without necessarily adjudicating in a decisive or systematic way, basic questions about environmental values that no scholar—indeed, no attentive reader—can ever again afford to ignore. If and when that happens, the claim made by Roger Payne for the force with which *Moby-Dick* dramatizes the mesmeric power that whales can exert over human minds and souls may actually start to hold for the power of environmental ethics, too: that they will "reintegrate at the point of origin of all the meridians of the imagination, its very pole, and there tie themselves forever into human consciousness by a kind of zenith knot."[16] If enough key people do, indeed, get knotted up in such a way, maybe we shall be able to save the world after all rather than sink like Captain Ahab.

In singling out this particular passage, I do not mean to leave the environmental reflection in *Moby-Dick* confined to reflection on species borderlines, cross-species analogies, and the like. On the contrary, other dimensions of Melvillian environmental imagination are arguably at least as important, whether from an artistic or an ecological standpoint. My analysis so far runs the risk, for example, of falling into a version of the traditional well-meaning protectionist shortsightedness of thinking about "species" without thinking "habitat" or "ecosystem." After the lecture presentation of the original version of this essay, a colleague in environmental science criticized my account of *Moby-Dick* as reflecting precisely this kind of myopia. Here the fault is partly with the novel, but mainly with my reduction of its environmental dimension to a single facet for the sake of illustration.

On the one hand, Melville had no real inkling of the prospect of eventual cetacean endangerment, of the late twentieth-century predicaments of large-scale oceanic pollution and depletion of whalestocks by high-tech weaponry and mega-sized factory ships. From his particular historical angle of vision, he had good reason to "account the whale immortal in his species, however perishable in his individuality."[17] Although he was quite aware—as this same passage elsewhere makes clear—of growing concern within the industry about the possible scarcity of sperm whales and other targeted species arising from the marked increased in the average length of whaling voyages from the beginning of the century, Melville's considered judgment that whales were growing warier rather than significantly rarer is borne out by the most authoritative study to date.[18]

On the other hand, *Moby-Dick* does show keen awareness of cetacean behavior in its larger ecological context: e.g., of sperm whales' global range and migration routes, of their favored locations at different seasons, of their social habits, of their food sources. To be sure, considered purely at the level of informational content, there is nothing especially deep or prescient about this store of knowledge. Any intelligent whaleman of the day could have equaled or beat Melville at his own game in this respect. What is especially striking, if not unique about the novel in this regard, is the sense of surging semi-chaotic mental energy and the aura of the magical with which it transfuses this material, as when "The Chart" chapter, which focuses on Ahab's strategy for tracking the white whale around the world, remarks how sperm whales swim ". . . .from one feeding-ground to another, . . . guided by some infallible instinct—say, rather, secret intelligence from the Deity—mostly . . . in veins, as they are called; continuing their way along a given ocean-line with such undeviating exactitude, that no ship ever sailed her course, by any chart, with one tithe of such marvelous precision."[19] Here again the text enriches environmental reflection by hedging assertion, in this case of the amazing accuracy of whale navigation, with a hypothetical equivocation as to the cause ("say, rather . . .") that elsewhere the novel mischievously calls into question. In this way, Melville distinguishes between the narrator's disposition to probe, query, theorize, and momentarily pontificate, but ultimately raise more questions than he answers versus Captain Ahab's will to achieve a precision equal to the sperm whale's own in the tracking of this particular creature—an obsession that Ishmael treats with a similar mixture of awe and bemusement. In this way *Moby-Dick* conveys a sense of the intricacies of whaling expertise and cetacean ecology as objects of great fascination and significance, ethically as well as metaphysically and aesthetically, while avoiding fixed ethical judgments.

At this point I can imagine an objection might arise to my choice of *Moby-Dick* as being too convenient an example of my broader claim. After all, this novel is a notoriously elusive and multifarious text, overstuffed with innumerable speculation of many kinds. Nor, until the last ten years or so, has it been thought of as part of "an environmental canon," much less as an "environmentalist" work. What if one were to choose as the preferred example, say, one of the other books mentioned earlier, where the historical track record of literary text acting as environmental(ist) provocation is much more clear cut? Let's turn, then, to Edward Abbey's contemporary eco-radical classic, *The Monkey Wrench Gang*.

To appearances at least, Abbey is a less complicated and more explicitly didactic writer than Melville. "I write to make a difference" is his opener to his last completed book, two weeks before his death.[20] Not only did *The Monkey Wrench Gang* help inspire Earth First!, Abbey became an early member, a rousing speaker at its inaugural meeting and others thereafter, and a vocal promoter of the kind of environmental sabotage, "ecotage" as he called it, that his novel describes. Abbey contributed the "Forward!" to movement founder Dave Foreman's how-to-do-it monkeywrenching manual, *Ecodefense* (1985), stressing that "never was such a book so needed." At least seemingly, Abbey makes his own environmental radicalism crystal clear: "If the wilderness is our true home, and if it is threatened with invasion, pillage and destruction—as it certainly is—then we have the right to defend that home, as we would our private rooms, by whatever means are necessary."[21] Yet it could also be argued that what makes *The Monkey Wrench Gang* a compelling book is not so much its ecodidacticism as any or all of the following: its action-packed plot culminating in an extended chase sequence that takes up the last four chapters, its Thoreauvian penchant for seriocomic over-the-top one-liner pronouncements ("our only hope is catastrophe"),[22] and the four colorful characters, or rather caricatures, that form the gang—Doc Sarvis, an Albuquerque surgeon who destroys billboards as a nighttime hobby to stave off midlife depression; Bonnie Abzug, his nurse, mistress, and later wife, a young

Jewish woman from the Bronx with a quick wit and an intellectual streak; Seldom Seen Smith, a Mormon whitewater rafting guide with three wives whose town has been buried by Glen Canyon Dam; and especially George Washington Hayduke, a foul-mouthed and unpredictable ex-Green Beret unhinged and alienated from techno-industrial civilization by his Vietnam years.

But these neat capsule descriptions threaten to normalize the novel's rambunctiousness and obscure the point I want to make about it. Let us again go to a particular passage, a campfire exchange during the rafting trip where the quartet first meet up with each other. Sarvis has just chided Hayduke for littering highways with beer cans.

> "Hell," Smith said. "I do it too. Any road I wasn't consulted about that I don't like, I litter. It's my religion."
>
> "Right," Hayduke said. "Litter the shit out of them."
>
> "Well, now," the doctor said. "I hadn't thought about that. Stockpile the stuff along the highways. Throw it out the window. Well . . . why not?"
>
> "Doc," said Hayduke, "it's liberation."
>
> The night. The stars. The river. Dr. Sarvis told his comrades about a great Englishman named Ned. Ned Ludd. They called him a lunatic but he saw the enemy clearly. Saw what was coming and acted directly. And about the wooden shoes, les sabots. The spanner in the works. Monkey business. The rebellion of the meek. Little old ladies in oaken clogs.
>
> "Do we know what we're doing and why?"
>
> "No."
>
> "Do we care?"
>
> "We'll work it all out as we go along. Let our practice form our doctrine, thus assuring precise theoretical coherence."[23]

Much rides on this passage, not only within this text but also historically, such that a case could be made for taking it with utmost seriousness. Within the novel, this is the moment when the gang first comes together. In Doc Sarvis' hastily summarized monologue are remembered for the first time and most conspicuously in the book the earliest precursors invoked both by these imaginary radicals and later by the Earth First! Movement the book helped bring into being—namely the Luddites and the French revolutionaries. The former supplied the name of the publishing house most closely associated with the movement (Tucson's Ned Ludd Books), the latter the neologism "ecotage," which like "monkeywrenching" this book popularized. On the other hand, parts of the passage are clearly meant to sound nutty and slapdash: the religion of littering, the insouciant "Do we know?" and "Do we care?" As for Doc Sarvis' reliability, even though he emerges as on balance the most thoughtful as well as best-educated member of the gang, the reader has already been put on notice that he can be a loony dreamer too. The "great Englishman named Ned" could just as easily be read as Sarvis' over-the-top enthusiasm rather than as an authorial endorsement. His pronouncement that putting practice before theory ensures "precise theoretical coherence" is clearly meant to sound like swagger, if not just amusingly absurd.

Then, too, as the passage suggests and the rest of the novel bears out, although the gang bonds and hangs together they never fully agree. They voice different, often shifting, motives and strategies, "stand for" different kinds of environmentalist commitment. Indeed their "commitments," such as they are, cannot be surgically extricated from the anger, impulse, and pranksterism that continually overtake them. Most importantly perhaps, Hayduke is shown clearly to differ from the rest in advocating violence against people as well as property. He always loses when it comes to a vote, and he is made to confess that he is a war-damaged mental case but he is also romanticized as the only one of the four who eludes capture, plea bargaining, and submission to surface respectability. Although Doc Sarvis firmly stipulates "no violence," and although this remains the group's—and Earth First!'s—"official" position, the novel thereby leaves the underlying question open, rather like the sweeping "by any means necessary" in Abbey's foreword to Foreman's ecodefense manual, which itself explicitly counsels forms of eco-subversion that will not harm humans.

All this goes to show that even a work of literature so obviously partial to environmental activism as *The Monkey Wrench Gang*, to the point that it has become a classic and even a template for latter-day environmental radicals, cannot be reduced to a single doctrine or line of argument. Abbey was justified in complaining that *The New York Times* review of the novel "misrepresents the book as a 'revolutionary' tract for the old 'New Left,'" with "no mention of the comedy, the wordplay, the wit, humor, and brilliance!"[24] That does not mean, however, that one should disbelieve Abbey's claim that he wrote to make a difference, only that the nature of the difference cannot be pinned down too precisely. Not only does art usually resist this kind of reduction—with rare exceptions on the order of

John Bunyan's allegorical narrative of his protagonist Christian's journey of salvation from the City of Destruction to the Celestial City—it is no less true, as Upton Sinclair found to his chagrin, that even a purposeful writer cannot guarantee the terms of his reception. In the last analysis the readers will have their way. You become your admirers, as the poet W. H. Auden put it in his great elegy on the death of fellow poet W. B. Yeats. It follows that the book which provokes a particular response cannot be held entirely responsible for it. Should we hold Henry Thoreau posthumously accountable for the thousands of life-altering experiments ventured in his name? Imagine for instance a meeting in the Elysian Fields between Thoreau and Christopher McCandless, the woefully underprepared suburban youth whose misdirected sortie into the Alaska wilderness a dozen years ago ended in death by starvation. In the copy of *Walden* McCandless took with him was highlighted the assertion, "No man ever followed his genius till it misled him. Though the result were bodily weakness, yet perhaps no one can say that the consequences were to be regretted, for this were a life in conformity to higher standards."[25] Would McCandless be justified in reproaching Thoreau for leading him down the primrose path? Thoreau might easily retort with this other pronouncement from *Walden*, "I would not have any one adopt my mode of living on any account; for, beside that before he has fairly learned it I may have found out another for myself, I... would have each one be very careful to find out and pursue his own way, and not his father's or his mother's or his neighbor's instead."[26]

The Auden poem, which I mentioned in the paragraph above, albeit in no sense an environmentalist text, sheds further light on the question of just how a creative writer might "make a difference." At first the poem makes what looks like a preemptively dismissive view of this prospect, as if to arc protectively within its own cocoon against the vulgar public: "poetry makes nothing happen: it survives/ In the valley of its saying where executives/ Would never want to tamper." But it ends oppositely, on a different mental wavelength entirely, with a call to poets to exert moral leadership in this dark time (with World War II looming): "In the prison of his days/ Teach the free man how to praise."[27] Why this apparent self-contradiction about art's social role? Clearly in order to press to the limit the distinction between an instrumental purpose and an inspirational one. The two are not quite so antithetical as Auden proposes; he, too, took a good thing too far. But unless one understands the distinction, so crucial to how creative imagination and proper critical practice both work, one is likely to fall far short of grasping the social impact art can make.

Edward Abbey was fortunate to have had an elegist in Earth First!'s founder, Dave Foreman, for whom Abbey's work had made a profound, life-changing difference and who was capable of expressing that difference in a way that did justice to the distinction Auden makes. In an affectionately Abbeyesque tribute written shortly after his death, Foreman remembers him as:

> . . . The Mudhead Kachina [the troublemaking prankster-figure] of the conservation movement, perhaps of the whole goddamned social change movement in this country. He was Coyote. Farting in polite company. Enraging pompous prudes, prigs, and twits. Goosing the True Believers. Pissing on what was politically correct.
>
> And thereby doing sacred work.
>
> Ed understood deeply the need for balance. He wrote, "Be as I am . . . a part-time crusader, a half-hearted zealot. . . . It is not enough to fight for the WEST; it is even more important to enjoy it." Whenever we are overworked and overwhelmed, whenever we lose our balance and our perspective, we need to read that wise advice from Abbey. [28]

Obviously this is not the only way environmental imagination might be thought to energize an environmentalist True Believer, precisely by not behaving in an instrumental way. The art of Thoreau, of Melville, of Sinclair, even of Carson, who in *Silent Spring* comes closest of the group to subordinating literary art to documentary polemic, would each require a somewhat different epitaph. But in each case the power of their books resides especially in their being something other, something more complexly haunting, than handbooks or directives: resources to energize the spirit, as Foreman puts it, when it is in danger of faltering—or rigidified by its own true believerhood.

NOTES

1. Histories of wetlands protection generally grant literature and the arts at least a bit part in the long campaign to transvalue and protect wetland spaces, e.g., Ann Vileisis, *Discovering the Unknown Landscape* (Washington, D.C.: Island Press, 1997), 150–151, 220. The fullest and most inventive study, albeit somewhat tendentious and uneven, of contemporary versus traditional artistic and critical valuation of wetlands is Rodney James Giblett, *Postmodern Wetlands: Culture, History, Sociology* (Edinburgh: Edinburgh University Press, 1996).
2. Ulrich Beck, "Politics in Risk Society," *Ecological Enlightenment: Essays on the Politics of the Risk Society*, trans. Mark A. Ritter (Atlantic Highlands, N.J.: Humanities Press, 1995), 14.
3. Matt Cartmill, *A View to a Death in the Morning: Hunting and Nature Through History* (Cambridge: Harvard University Press, 1993), 161–188.
4. On the rhetorical devices and power of *Silent Spring*, see especially Craig Waddell, ed., *And No Birds Sing: Rhetorical Analyses of Rachel Carson's "Silent Spring"* (Carbondale: Southern Illinois University Press, 2000). Priscilla Coit Murphy in *What a Book Can Do: The Publication and Reception of "Silent Spring"* (Amherst: University of Massachusetts Press, 2005) makes clear both that scientific credibility was a top priority for Carson, her publishers, and her agent but also that her prior reputation and literary skills were crucial to the promotion of the book. For example, Carson's *New Yorker* editor, William Shawn, congratulated Carson on having made her project "literature full of beauty and loveliness," and saw to it that the manuscript was revised so as to backload the technical information to a greater extent and make the manuscript as stylistically compelling as possible for the nonspecialist reader (Murphy, 63).
5. The quotation is from the first full-length scholarly history: Martha F. Lee, *Earth First! Environmental Apocalypse* (Syracuse, N.Y.: Syracuse University Press, 1995), 65—a judgment upon which movement watchers generally agree. "Abbey's fictional vision inspired non-fictional action." Stephen Best and Anthony J. Nocalla, "Introduction," *Terrorists or Freedom Fighters? Reflections on the Liberation of Animals*, ed. Best and Nocella (New York: Lantern Books, 2004), 59. The three previous cases just cited are all (even more) familiar and fully documented.
6. Henry David Thoreau, *Walden* (1854), ed. J. Lyndon Shanley (Princeton, N.J.: Princeton University Press, 1971), 107, 74. Edward Abbey, a devoted though also a critical Thoreauvian, recycles this aphorism in a typical George Hayduke hyperbole: "'When I see somebody coming to do *me* good, . . . I reach for my revolver'" in *The Monkey Wrench Gang*, (New York: Avon, 1975), 109.
7. For representative examples of the former, see John Elder, *Reading the Mountains of Home* (Cambridge: Harvard University Press, 1998); and Ian Marshall, *Peak Experiences: Walking Meditations on Literature, Nature, and Need* (Charlottesville: University Press of Virginia, 2003). For representative examples of the latter, see Joni Adamson, *American Indian Literature, Environmental Justice, and Ecocriticism: The Middle Place* (Tucson: University of Arizona Press, 2001);

and a number of the contributions to Adamson, Mei Mei Evans, and Rachel Stein, ed., *The Environmental Justice Reader: Politics, Poetics, and Pedagogy* (Tucson: University of Arizona Press, 2002). See my *The Future of Environmental Criticism* (Malden MA: Blackwell, 2005), 17–28 and passim, for further discussion of the relationship between the two "waves" (my coinage).

8. First quotation from Upton Sinclair, "What Life Means to Me" (1906), reprint *The Jungle*, ed. Clare Virginia Eby (New York: Norton, 2002), 351. Sinclair here declares that *Uncle Tom's Cabin* was "a model of what I wished to do" (350), although my second quotation is from his friend and fellow-writer Jack London's review of *The Jungle* reprinted in that same edition (483).
9. For the biographical record, see Anthony Arthur, *Radical Innocent: Upton Sinclair* (New York: Random House, 2006), 80–84, 322, pointing out that Roosevelt came to think Sinclair a crackpot.
10. Edward O. Wilson, "A Letter to Thoreau," *The Future of Life* (New York: Knopf, 2002), xi–xxiv.
11. See particularly the introductory chapter to my *Writing for an Endangered World: Literature, Culture, and Environment in the United States and Beyond* (Cambridge: Harvard University Press, 2001).
12. My discussion draws to some extent on my previous discussion of this novel in *Writing for an Endangered World*, 205–214.
13. Herman Melville, *Moby-Dick*, ed. Hershel Parker and Harrison Hayford, 2nd ed. (New York: Norton, 2002), 242.
14. Roger Payne, *Among Whales* (New York: Delta, 1995), 325.
15. Some relevant contemporary discussions of different approaches and persuasions include Elizabeth Schultz, "Melville's Environmental Vision in *Moby-Dick*," *Interdisciplinary Studies in Literature and Environment* 7 (2000): 97–113; Eric Wilson, "Melville, Darwin, and the Great Chain of Being," *Studies in American Fiction* 28 (2000): 131–150; William Howarth, "Earth Islands: Darwin and Melville in the Galapagos," *Iowa Review* 30 (2000): 95–113; and Lilian P. Carswell, "Telling the Truth about Animals: Epistemology, Ethics, and Animal Minds in Melville, Darwin, Saunders, and London," PhD diss., Columbia University, 2004.
16. Payne, *Among Whales*, 325.
17. Melville, *Moby-Dick*, 354.
18. Lance E. Davis, Robert E. Gallman, and Karen Gleiter, *In Pursuit of Leviathan: Technology, Institutions, Productivity, and Profits in American Whaling, 1816–1906* (Chicago: University of Chicago Press, 1997), 131–149.
19. Melville, *Moby-Dick*, 167.
20. Edward Abbey, *A Voice Crying in the Wilderness: Notes from a Secret Journal* (New York: St. Martin's, 1989), v.
21. Edward Abbey, "Forward!," *Ecodefense: A Field Guide to Monkeywrenching*, 2nd ed., ed. Dave Foreman and Bill Haywood (Tucson, Ariz.: Ned Ludd Books, 1987), 9, 8.
22. Abbey, *Money Wrench Gang*, 41—this from the thoughtstream of one of the

four main characters, Doc Sarvis, of whom more below.

23. Abbey, *Monkey Wrench Gang,* 65.
24. *Confessions of a Barbarian: Selections from the Journals of Edward Abbey,* 1951–1989, ed. David Petersen (Boston: Little, Brown, 1994), 245.
25. Thoreau, *Walden,* reported in Jon Krakauer, *Into the Wild* (New York: Villard, 1996), 47.
26. Thoreau, *Walden,* 71.
27. W. H. Auden, "In Memory of W. B. Yeats (d. Jan. 1939)," *Collected Poetry* (New York: Random House, 1945), 50, 51.
28. Dave Foreman, *Confessions of an Eco-Warrior* (New York: Harmony Books, 1991), 174–175. Bracketed explanation added; ellipses in the Abbey quote are from the original.

Nature, Liberty, and Equality[1]

Donald Worster

The struggle to protect wild nature goes on all over the planet, from Brazil to Zimbabwe, but we still have not explained fully why people care. One set of explanations derives from examples like Ashoka, the ancient ruler of India (third century BCE), who set aside the world's first wildlife preserve after converting to Buddhism and its doctrine of *ahimsa*, or non-violence toward all living things. His example suggests either that religion has been the driving force or that powerful elites deserve credit for protecting the natural world. Both explanations can claim a degree of truth. But the most active nations in nature protection have not been especially devoted to Buddha or other traditional religions, while most elites, from emperors to corporate executives, have been destructive of or indifferent to nature.[2]

Ordinary people, on the other hand, whether conventionally religious or not, have often found delight in the smell of a forest or the sight of a wild antelope. The protection of nature owes much to them too. They have been far more important than historians have commonly acknowledged. We have not fully appreciated how much the protection of nature owes to the rise of modern liberal, democratic ideals and to the support of millions of ordinary people around the world.

The role of democracy in promoting nature protection becomes apparent when we examine where most of that protection has occurred in the modern world. Overwhelmingly, it has taken place within nations that profess democratic principles, cherish human rights, and allow freedom of speech and dissent from established elite opinion. Wherever open, egalitarian societies have taken root, protection has spread rapidly; conversely, it has failed wherever it has been confronted by powerful technocrats, politburos, and other religious or political forms of authoritarianism.

Fortunately for a world undergoing a continuing democratic revolution, there is plenty of wild nature left to protect. In 1989 a reconnaissance survey found that 48 million square kilometers of the planet qualified as wilderness, or about a third of the total land surface.[3] (Forty-eight million square kilometers is equivalent to twelve billion acres, an expanse larger than the Western Hemisphere.) Fifteen million of those square kilometers are in Antarctica and Greenland—vast white wildernesses of ice.[4] Much of the earth's surface in the higher latitudes is remarkably wild, as are much of the world's deserts and tropical rainforests and virtually all of the oceans, where until very recently there have been few traces of human impact.

Traditionalists might insist that wilderness must mean forested mountains, not glaciers or ice sheets or oceans, but that would be a highly arbitrary definition. Wilderness, according to the survey, does not refer to a particular kind of biome; it can include any sort of nature that shows little sign of active human settlement or commodity production, whether forests, grasslands, deserts, polar caps, volcanic plains, or lakes and seas.

The 1989 survey looked for areas larger than 400,000 hectares (one million acres) that lacked any "permanent human settlements or roads," lands that were "not regularly cultivated nor heavily and continuously grazed" but that might have been "lightly used and occupied by indigenous peoples at various times who practiced traditional subsistence styles of life." Wilderness purists might not like the looseness of that standard; for some, a single tissue can spoil a place, or a solitary fisherman's hook, or a lone donkey track. For those *opposed* to any strict protection, on the other hand, even the light passage of a primitive tribe through the landscape should disqualify it as wilderness and, without much analysis, they want to put it in the category of a "well-used" or "managed" place, open to exploitation. Neither kind of absolutism will do; neither reflects the flexible, pragmatic definitions people have historically used or the inescapable relativity of the term.[5]

The United States, despite its persistent frontier image, ranked low on the list of wilderness-rich countries—down at number sixteen, with only 440,580 square kilometers (109 million acres), or 4.7 percent of its total area. Higher on the list were Russia, Canada, Australia, Brazil, Sudan, and

Algeria. Several heavily populated countries were surprisingly high on the list, including China, India, Laos, Mexico, and Iraq. China, for instance, despite its one billion-plus population, still had 22 percent of its territory in a wild state, a far higher percentage than the U.S.

A survey that focuses only on huge, million-acre parcels of land does not exhaust the possibilities of wild places on the Earth. There are many places under that size that might qualify as wild—a mere ten thousand or a hundred thousand acres in extent. And then there are all those smaller, even tiny, patches of wildness that lurk on the edges of our cities, farms, and backyards and that may be wonderfully rich in diversity and high in aesthetic and spiritual value.

Where the U.S. stands high among nations, where it might even be called exceptional, is not so much in the size of its wild lands as in its long history of activism in protecting them. The U.S. was the first nation to create a national park (in 1864 or 1872, depending on whether one grants priority to Yosemite or Yellowstone), the first to set up a full-blown "wilderness preservation system" (1964), and the first to pass an endangered species act (1973).[6] That historic leadership role seems to have come abruptly to an end, following the conclusion of the Clinton presidency, which in one magnificent moment declared more than 60 million acres of U.S. forest lands to be forever free of roads. That ruling was quickly suspended by the second Bush administration. In recent years, the cause of wild-lands protection has been rejected on the political right for stifling private enterprise and has been much criticized by some on the political left for detracting attention from issues of social justice.[7] As a consequence, leadership in nature protection has passed to other nations, some of which are the older democracies while others are relatively younger nations still struggling to transfer more power to the people and to make nature preservation part of their culture.

That shift in leadership was noticeable at the 1992 Earth Summit in Rio de Janeiro when the U.S. took a back seat as over a hundred nations agreed that every country should protect at least 12 percent of its land base from economic use. Not every nation voting at that meeting was a full-fledged democracy, but the decision was one that reflected a democratic process of open discussion and global representation. It was animated by an egalitarian purpose—to protect the beauty, health, and integrity of nature for the sake of future human generations and to recognize a moral obligation to save other forms of life from extinction.

Preservationists all over the world have agreed on a common program to set up protective zones where farming, logging, mining, town building, wildlife poaching, or the dumping of wastes is prohibited or severely restricted. They have represented a wide array of ethnic groups as well as languages. The Nordic countries, for example, have produced plenty of activists and can boast some of the most carefully protected wild lands in the world. Thousands of miles away, and sharply contrasting in many ways, is Costa Rica, which has protected 28 percent of its territory from development—11 percent in national parks, 4 percent in indigenous reserves, and 13 percent in a miscellaneous series of biological reserves, national forests, national monuments, and national wildlife refuges.[8] The spectacular diversity of its flora and fauna, the stunning beauty of its mountain ranges, exuberant wet and dry forests, and broad saltwater beaches, have given rise to one of the world's most conservation-minded societies. Next door, Panama in its post-Noriega period is moving toward a similar policy of large-scale, vigorous nature protection. What joins those two Central American countries to Norway, Finland, or Sweden, or joins any of them to New Zealand and its great protected wilderness of Milford Sound? Why are many other nations so backward in preserving wild places—Russia, for example, or Guatemala or Thailand?

The conventional answer is that preserving nature appeals only to affluent people whose stomachs are full and is never important to the poor or the aspiring. At the extremes this seems to be true; desperately hungry men and women are not likely to think much about wilderness or, indeed, think much about many other large issues at the national or global scale. Such an economic explanation is too simplistic and reductive to be dependable. Income alone does not work very well *within* societies in predicting which citizens care about preservation and which do not; it cannot explain why some oil executives care while plenty of others do not, nor, on the other hand, why some pensioners care while others among them do not.

Nor does a simple economic explanation work at the international level. According to World Bank data from 2006, Norway stands second in the world in gross national income per capita (U.S. $71,240); Finland, sixteenth ($41,360); New Zealand, thirty-seventh ($26,750); Panama, eighty-ninth ($5,000); and Costa Rica, ninetieth ($4,980).[9] Huge differences in wealth, yet all are active countries in nature awareness and preservation. Furthermore, within the most abysmally poor countries, where there may be little or no organized movement for preservation, many people care deeply about wildlife and unspoiled natural beauty.

A more reliable indicator of whether nations become active in preserving wild places is the state of personal freedom, the degree of social equality, and the sanctity of human rights. Far more than religious or ethnic identity or gross national product, the quality of nature protection seems to correlate with the quality of democracy. Countries where there is a more equitable distribution of economic opportunity, a low level of militarism, high levels of literacy, greater racial and gender equality, free and competitive elections, and tolerance of dissent tend to set aside significant pieces of nature for protection from economic development. Or, if they are too densely settled for that to be a realistic possibility at home, they work to do the same internationally—as Denmark has recently done in setting aside much of Greenland as the world's largest national park. Why that should be so, why liberal democracy should correlate to wild-land protection, is a question that has never been fully explored, although it is of the utmost importance to the future of life on Earth.

The history and meaning of liberal democracy is an old and complicated subject. We have come to realize that it refers to more than the superficial mechanics of political modernization—elections, parliaments, or governmental checks and balances—that liberal democracy is founded on a pair of intertwined cultural ideals: personal liberty and social equality. The greatest proponent of those ideals was Jean Jacques Rousseau (1712–78) who insisted that one ideal could not exist without the other. A government that promised equality to its citizens would never deliver without the constant pressure of free, critical, and dissenting opinion. Equality needed liberty, and liberty needed equality. That linkage has often come under challenge by those who want to promote one ideal but not the other: for example, political philosophers like Alexis de Tocqueville (1805–59), author of *Democracy in America*, who preferred liberty over equality, or politicians like Mao Zedong (1893–1976, communist dictator of China from 1949 until his death), who dismissed liberty in pursuit of the classless state. But the critics have not succeeded in splitting them apart. The two ideals have not always been easy to reconcile, but together they have worked to change the course of Western history and, increasingly, to change the dynamics of non-Western societies as well.

Much has been written on how that pair of ideals has revolutionized human relations but rather less on how they have affected people's relation to nature.[10] Their environmental impact has been little short of revolutionary too. Old notions that humans have been created specially in the image of God or that they have been given dominion over all other forms of life or that they can draw a rigid line around their own liberty or equality, making those ideals exclusive to *Homo sapiens*, have proved unsustainable. Nature, in fact, has become the patron and partner of liberal democracy. It has even come to be seen as the source of human liberation, a place of freedom and of equality, and therefore worthy of respect, protection, and even worship.

"I wish to speak a word for Nature," declared Henry David Thoreau in 1862, "for absolute freedom and wildness, as contrasted with a freedom and culture merely civil."[11] Going into wild country, as Thoreau advocated, experiencing places free of human domination, became a means of freeing oneself from the hand of convention or authority. Social deference faded in the wilderness. Economic rank did not matter so much. Money was not needed to survive there. Nature offered a home to the dissident mind, the rebellious child, the outlaw, the runaway slave, the soldier who refused to fight, and (by the late nineteenth century) the woman who went mountain climbing to show her strength and independence.[12]

A move toward greater equality among species became irresistible too, giving rise to animal rights, wildlife refuges, and even Darwin's theory of evolution, which joined humans and other forms of life into a common family. Plants and animals came to be valued for more than their potential for domestication, their fitness for pulling a wagon or yielding a crop; wild species came to be admired for surviving on their own, independent of human purposes. They were seen to form their own communities. They were not inferior versions of ourselves, but beings created by God or evolving by natural processes for their own sake. They were, as John Muir argued, "earth-born companions and our fellow mortals."[13]

Nature in the wake of liberal democracy also became the basis of a new (or rediscovered) religion, a fathomless source of spirituality, complementary to or independent of traditional religion. Woods, mountains, or prairies became divine texts in which one could find answers to life's ultimate questions, without the mediation of church authorities or theologians. Protestants in Western Europe led the way to this new religion by challenging the entrenched hegemony of the Pope and Roman Catholic Church and by insisting that every individual has a right and duty to read the Holy Bible for her/himself. They, and particularly groups like the Quakers and Presbyterians, opened a challenge to hierarchical religion that in turn they had trouble controlling within their own denominational walls. Any written Bible or testament came to be seen as a man-made artifact full of human frailties and limitations, inferior to the outdoors as a source of inspiration. Nature drew people away from all established creeds and faiths. In the presence of nature, the rising liberal spirit of the nineteenth and twentieth centuries found a new source of guidance accessible to any individual.

One of the great pioneers of that new religion of nature was none other than Rousseau. In 1762 both the French and Swiss government threatened him with arrest for being a dangerous heretic, a radical, and anti-Christian. Seeking refuge from the authorities, he travelled to St. Pierre's Island in Lake Biel near Bern, Switzerland and immersed himself, body and mind, in the wholeness of nature. His memoir *The Reveries of the Solitary Walker* tells about finding a subversive source of spiritual insight.

> The earth, in the harmony of the three realms [mineral, plant, animal], offers man a spectacle filled with life, interest, and charm—the only spectacle in the world of which his eyes and his heart never weary. The more sensitive a soul a contemplator has, the more he gives himself up to the ecstasies this harmony arouses in him. A sweet and deep reverie takes possession of his senses then, and through a delightful intoxication he loses himself in the immensity of this beautiful system with which he feels himself one.[14]

Others felt this call of nature, from William Wordsworth and Johann Goethe down to Rachel Carson and Robert Marshall. Whatever their national or religious roots, they have broken free from orthodoxy and found in nature part or all of what they needed to feed their spiritual hunger.

If the nature protection movement has been part and parcel of liberal democracy, influenced deeply by the ideals of liberty and equality, then we should not expect to find that movement blooming in places where repressive authority and inequality stand in the way. We should not expect a preservation ethic to flourish in a man like Anastasio Somoza, the dictatorial president of Nicaragua during the 1940s and 50s; or Colonel Joseph-Désiré Mobutu, the kleptocratic strong man of Zaire until deposed in 1997; or in such totalitarians as Mao, Stalin, or Pol Pot. We should not be surprised that wild lands are not attractive to military juntas, theocracies, patriarchs, or slave regimes. Nature in its wilder state is a threat to such authoritarian minds. It is where danger lurks, threatening always to erupt and bring down their vulnerable edifices of control.

We should expect, on the other hand, that nations in the forefront of nature preservation would be those influenced by ideals of liberty and equality, and indeed that is so: Costa Rica and Panama in Central America; New Zealand and Australia in the South Pacific; the United States and Canada in North America; Norway, Scotland, and others in Europe; and a new Bulgaria, Chile, India, or Zambia.

Liberal democracies are, of course, more than expressions of cultural ideals. They are also systems of governance. To do that work they must pass laws and regulations. In doing so they must infringe on the liberty of some citizens in order to protect the liberties of others, or to protect the spiritual values of wilderness or the rights of other species to survive. This rule making can lead to charges of injustice, and sometimes the charges are justified. Liberal democracies, in their making of laws and regulations, have not been free of class, gender, or racial bias, or always respectful of differences of opinion. They are imperfect creations. Tocqueville rightly warned about some of their shortcomings: a tendency toward tyranny of the majority exercised over minorities, a tendency to glorify greed (under the doctrine of economic liberalism), and a susceptibility to elites gaining power through the free and ruthless accumulation of money.[15] Those who demand their own freedom can be quick to deny it to others. Despite Rousseau's confidence that virtue must always flourish where liberty and equality together flourish, history shows a more complicated picture: liberal democracies that display hypocrisy along with virtue, conflict as well as cooperation, and bigotry of all sorts.

Similarly, the record of liberal democracies in protecting nature has often been flawed by narrow self-interest. The pursuit of liberty has at times meant the freedom to invade and exploit the natural world for personal gain. The pursuit of equality, for all of its positive appeal, has often led to environmental destruction; it has been one of the driving forces behind modern consumer culture, which promises everyone a more abundant material life and endless economic growth, regardless of the ecological consequences. Here again are contradictions difficult to resolve, impossible to avoid. Those contradictions have driven much of modern history. It is precisely because of them—the tension between liberty and equality, between present and future generations' claims on the earth, and between human rights and nature's right to exist—that liberal democracies do not represent some ultimate or finished victory. They are not, at least in their current forms, the "end of history."[16]

The most serious challenges facing conservationists are those regimes that have never been touched by or are falling back from liberal democratic ideals. They are many, and they control the destiny of much of the remaining wilderness on the planet. Some are still locked in repressive attitudes that allow no dissent from orthodoxy, no openness to new ideas or research, and no respect for the other-than-human world. Then there is the challenge of nations where liberal democracy is weakening or failing, as authoritarian forces within them gain strength. They too are not hard to find: Look for imperial-scale military budgets, social intolerance, education giving way to indoctrination, oil drilling in the last wild places, and dark warnings against "pagan" heresies.

Perhaps this is the way that the dream of liberal democracy self-destructs: in the quest for freedom and equality people may want to devour the earth rather than save it, and in devouring the earth they may lose the freedom they thought they were getting. They may end up as slaves to their own appetites, living in fearful bondage to whatever ideas or forces will offer them security. They may, in the words of Eric Fromm, seek "escape from freedom."[17] That seems to describe accurately the current mood of many Americans and others around the world.

But the historic association of nature protection with the spread of human liberty and human equality is a strong idea and a strong force too. That association has now reached into almost every corner of this imperiled planet. It may prove powerful and decisive, with a long future ahead for a global ethic of conservation.

NOTES

1. This essay "Nature, Liberty, and Equality" was published in slightly different form in the volume edited by Michael Lewis, *American Wilderness: A New History* (New York: Oxford University Press, 2007), 263–272. Used by permission of the author and Oxford University Press.
2. The role of elites in preserving nature cannot be denied, often played to secure good hunting or exotic travel for themselves. However, people of wealth and power may also be moved by genuine concern for less affluent citizens or for other species. These conflicts are well discussed in Jane Carruthers' work on South Africa, *The Kruger National Park: A Social and Political History* (Pietermaritzburg, South Africa: Natal University Press, 1995).
3. J. Michael McCloskey and Heather Spalding, "A Reconnaissance-Level Inventory of the Amount of Wilderness Remaining in the World," *Ambio* 18 (1989): 221–27. The survey did not include the 70 percent of the earth's surface covered by oceans, most of which is hardly explored in depth, let alone domesticated. The survey was too early to consider the effects of anthropogenic climate change on pristine environments, including the melting of ice sheets.
4. Since that survey the Danish government has set aside most of Greenland as the world's largest national park, covering 972,000 square kilometers. According to the United Nations Environment Program, approximately 19 million square kilometers, an area the size of Canada and the United States combined, have been given some protection globally. Eleven percent of that total, or two million square kilometers, has been placed under "strict" protection as a wilderness or nature reserve. See Table 1, "2003 United Nations List of Protected Areas," http://www.unep-wcmc.org.
5. A more difficult challenge for any definition of wilderness comes from the new potential for anthropogenic change in the global climate system. But even if we grant such change as scientific fact, it does not follow that we should now call every place on earth a "cultural landscape." A cultural landscape has been deliberately shaped by ideas and values, while global warming, anthropogenic or not, is as unwitting and unpredictable as a meteor hitting the earth. Moreover, a land left free of ice by global warming may still be "wild" if it is unsettled by human population or unexploited for commodities.
6. In some places protection commenced almost as long ago as in the United States: in 1887, for example, the Maori leader Te Heuheu Tukino IV gave the austere volcanic peak region of the North Island, now the Tongariro National Park, to the nation of New Zealand. With additional acreage added, the area was incorporated as a national park in 1894. Ten years later that country made the fjord lands of the South Island off-limits to economic development. See Paul Star and Lynne Lochhead, "Children of the Burnt Bush: New Zealanders and the Indigenous Remnant, 1880–1930," in *Environmental Histories of New Zealand*, ed. Eric Pawson and Tom Brooking (Melbourne, Australia: Oxford University Press, 2002), 123–27.

7. For a critique of this debunking spirit among historians see my essay, "The Wilderness of History," *Wild Earth* 7 (Fall 1997): 9–13.
8. Sterling Evans, *The Green Republic: A Conservation History of Costa Rica* (Austin: University of Texas Press, 1999), 7–8.
9. See the comparative tables on per capita gross national income (GNI) at the World Bank website: http://www.worldbank.org/data/quickreference/quickref.html.
10. An exception to this observation is Roderick Nash's, *The Rights of Nature: A History of Environmental Ethics* (Madison: University of Wisconsin Press, 1989). Although focused mainly on liberal democratic ideals within the United States, Nash does include such figures as the Norwegian Arne Naess, founder of the Deep Ecology movement, whose ideas seem profoundly indebted to Rousseau, William Wordsworth, and other early modern thinkers.
11. Henry David Thoreau, "Walking," *Atlantic Monthly* 9 (June 1862): 657.
12. See, for example, Susan R. Schrepfer, *Nature's Altars: Mountains, Gender, and American Environmentalism* (Lawrence: University Press of Kansas, 2005).
13. John Muir, *A Thousand Mile Walk to the Gulf* (Boston: Houghton Mifflin, 1916), 139.
14. Rousseau, *The Reveries of the Solitary Walker*, in *The Collected Writings of Jean-Jacques Rousseau*, ed. Christopher Kelly and trans. Charles Butterworth, Vol. 8 (Hanover, N.H.: University Press of New England, 2000), 59.
15. See Alexis de Tocqueville, *Democracy in America*, 2 vols. 1835–1840. The American Studies Program of the University of Virginia has a website devoted to the book, including the full translated text at http://xroads.virginia.edu/~HYPER/DETOC/home.html.
16. The environmental movement has laid bare those tensions within liberal democracy, although surprisingly it gets little credit for doing so in Francis Fukuyama's *The End of History and the Last Man* (New York: Free Press, 1992).
17. See Erich Fromm's influential book of that title, *Escape from Freedom* (New York: Holt, Rinehart and Winston, 1941).

Touching the Depths of Things: Cultivating Nature in East Asia

Mary Evelyn Tucker

As a historian of religion, especially East Asian religions, I am interested in understanding what these traditions have said in their own cultural and historical contexts, as well as what they say to us in our times. I acknowledge the concerns of those who argue for the incommensurate nature of cultural differences that are involved in studying other religions. This is why we rely on the rigorous scholarly work of historians of religions and theologians as well as thinkers in literature, art, history, and philosophy. At the same time, I observe, along with many others, that these traditions have significant ideas regarding who we are as humans and our relationship to nature, which transcend the particularity of cultural or historical contexts.

Indeed, it is the case that the traditions of Confucianism, Daoism, and Buddhism are being explored in East Asia as a source of environmental ethics. These investigations are taking place at the highest level of academic research in the Chinese Academy of Social Sciences in Beijing, in the Academia Sinica in Taiwan, and in the Academy of Korean Studies outside of Seoul. In addition, Pan Yue, the Vice Minister of the Environment in China, has expressed keen interest in drawing on traditional values to support environmental protection.

Confucianism: Resources for Ecological Perspectives and Ethics

In this spirit of acknowledging differences of time and circumstances yet valuing the efficacy of ideas transcending such particularity, I turn to my own field of Confucian studies to explore the work of Wang Yangming (1472–1529), a leading scholar, statesman, and soldier in the Neo-Confucian tradition of China. The recent book *Fifty Key Thinkers on the Environment* includes only six people who are not Western thinkers—from India, Buddha, Gandhi, and Vandana Shiva; from China, Chuang Tzu and Wang Yangming; and from Japan, Basho.[1] I cannot help wondering what this book will look like fifty years from now as we learn more about individuals beyond the West who will expand our environmental thinking from out of the humanistic legacy of world cultures that we are inheriting in all their rich diversity.

Wang Yangming is one of these individuals. Along with Chu Hsi (1130–1200), Wang was the most influential thinker in the Chinese Neo-Confucian tradition. His ideas spread across East Asia and endured in both Korea and Japan for some five hundred years. In Japan, Confucians interested in implementing humane government relied on his teachings as did the reformers in the Meiji Restoration in the nineteenth century. Seen as controversial by some more orthodox Confucians, he nonetheless has made important contributions to humanistic thought and praxis that warrant consideration today for environmental ethics—both for East Asia and for the planet as a whole.

The three pillars of his thought on epistemology, ethics, and cosmology are an emphasis on empathetic knowing, embodied acting, and compassionate living in the world. Empathetic knowing is affirming human subjectivity as a primary way of apprehending the nature of things. Embodied acting is unifying knowledge and action. Compassionate living is embracing a common kinship with the larger community of life. These ideas of Wang include both humanistic and ecological values. They can be more fully appreciated as situated within the broader context of Confucian and Neo-Confucian thought itself.

The Confucian tradition is remarkably rich and diverse from its early classical articulation by Confucius (551–479 BCE) and Mencius (372–289 BCE), to its theories of cosmological and correlative correspondences in the Han period (206 BCE–220 CE), through to its more metaphysical expressions in the Neo-Confucian synthesis of the Sung, Yuan, and Ming dynasties (tenth through seventeenth centuries).[2] Its humanistic values had a strong and lasting appeal, spreading across East Asia to Korea and Japan and into Southeast Asia to Vietnam and Singapore. Its influence continues in the present with new Confucian philosophers working out of Hong Kong, Taiwan, and Singapore along with Tu Weiming at Harvard.

Although the tradition is not well known in the West, the humanistic and ecological perspectives of Confucian and Neo-Confucian thinkers merit our attention. This tradition has dealt for 2,500 years with pressing questions of how to create sustainable societies supported by a politics that promotes the common good. Within this ancient Confucian ideal of establishing humane government, they encouraged the equitable distribution of goods and provided the conditions for agriculture and commerce to flourish. Against formidable odds and despite frequent failures they struggled with how to sustain effective political and social institutions. To staff these institutions they created the oldest meritocracy in the world in which government appointments were based on civil service examinations that drew on the values of the Confucian classics.

All leaders, even the emperor himself, were called to the task of ongoing moral cultivation. In this spirit Confucian education was a means of encouraging self-cultivation of the literati so they could contribute to the well-being of the larger society and body politic. Thus, Confucians promoted education, fostered printing, created immense libraries, established academies, and built universities. They also encouraged the arts, celebrated nature in painting and poetry, kept detailed historical records, and wrote lasting interpretive histories. While Confucianism, like all religious and philosophical traditions, at times became distant from its ideals and was used for autocratic ends, it nonetheless was a dynamic and unifying force for one of the world's oldest continuing civilizations.

Confucianism's promotion of broad humanistic education in the family and in schools is one of the reasons for its continuity into the present. The aim of such education and self-cultivation was to realize one's innate good nature and ultimately to achieve sagehood. This is what Wang Yangming prized above all—beyond fame, position, or power. Such cultivation of the person was based on a firm belief in the goodness of human nature as well as an understanding of the inherent unity of thinking and feeling. The Chinese character for mind implies both functions and is often translated as "heart-mind" or "mind and heart." The human in the Confucian context is considered to be the heart and mind of the universe, which is described as Heaven and Earth.

Another distinguishing feature of Confucianism is its recognition that the basis of a sustainable society and civilization is a healthy agricultural system. Thus Confucian governments built complex irrigation works as well as regional granaries for the storage of rice. Moreover, in the capital city, the emperor supported the development of farmers' almanacs and also performed rituals at the altar of Earth for the planting and harvesting of rice. All of this was done with detailed attention to seasonal changes and the appropriate cultivation of nature.

In East Asia, then, cultivating nature was a primary concern of Confucians. This cultivation embraced both the natural and human worlds, namely agricultural cultivation of the land for producing food and moral cultivation of the human for creating virtue. In essence these were seen as mutually related activities, for the human was not viewed as an isolated entity but as part of the larger community of life—social, natural, cosmological. Tu Weiming has called this an anthropocosmic worldview in contrast to the more anthropocentric emphasis of the Western religious and philosophical traditions.[3] This anthropocosmic worldview is distinguished by a naturalistic cosmology and a transformative ethics.

Naturalistic Cosmology

This naturalistic cosmology of Neo-Confucianism is characterized by an organic holism and a dynamic vitalism.[4] The organic holism of Neo-Confucianism refers to the fact that the universe is viewed as a vast integrated unit, not as discrete mechanistic parts. Nature is seen as unified, interconnected, and interpenetrating, constantly relating microcosm and macrocosm. This interconnectedness is already present in the early Confucian tradition in the *I Ching (Book of Changes)* and in the Han-period cosmological correspondences of the elements with seasons, directions, colors, and even virtues. Cheng Chung-ying has described the organic naturalism of Confucian cosmology as characterized by "natural spontaneity" and "inclusive humananness" in contrast to the emphasis on rationality and transcendence in Western thought.[5]

This naturalistic cosmology based on organic holism is distinguished by the view that there is no Creator God; rather the universe is considered to be a self-generating, organismic process.[6] Neo-Confucians are traditionally concerned less with theories of origin or with concepts of a personal God than with what they perceive to be the ongoing reality of this generative, interrelated universe. This interconnected quality has been described by Tu Weiming as a "continuity of being."[7] This implies a kind of great chain of being, linking inorganic, organic, and human life-forms. For the Neo-Confucians this linkage is based on the understanding that all life is constituted of *ch'i*, the material force or psychophysical dimension of the universe. This is seen as the unifying element of the cosmos and creates the basis for a profound reciprocity between humans and the natural world.

The second important characteristic of Neo-Confucian cosmology is its quality of dynamic vitalism inherent in *ch'i* (material force). Material force is the substance of life and the basis for the continuing process of flux and fecundity in the universe. The term *sheng sheng* (production and reproduction) is used in Neo-Confucian texts to illustrate the ongoing creativity and renewal of nature. Furthermore, *sheng sheng* reflects an ecological awareness that change is the basis for the interaction of life systems—mineral, vegetable, animal, and human. And finally, *sheng sheng* celebrates transformation as the clearest expression of the dynamic processes of life with which humans should harmonize their own actions. In essence, humans are urged to "model themselves on the ceaseless vitality of the cosmic processes."[8] This approach to self-cultivation is an important key to Neo-Confucian thought, for a cosmology of holism and vitalism provides a metaphysical basis on which an integral morality of harmonizing with change is developed.[9]

The extended discussions of the relationship of *li* (principle) to *ch'i* (material force) in Neo-Confucianism can be seen as part of the effort to articulate continuity and order in the midst of change.[10] The term "principle" is used to name the pattern amidst flux that provides a means of establishing harmony. This is in contrast to a Buddhist understanding of attachment to change as a source of suffering.

The Transformative Ethics of Self-Cultivation

For the Neo-Confucians the idea of self-cultivation implies, then, a "creative transformation"[11] such that humans form a triad with Heaven and Earth. This triad underlies the assumption of our interconnectedness to all reality; experiencing that connection is an overriding goal of self-cultivation. Interrelatedness is thus both a given and an achievement for humans. Through the deepening of this sense of basic connection, humans may participate fully in the transformative and fecund powers of the universe. In doing so humans are able to touch the depths of things.

In cultivating their moral nature within this triad, then, humans are entering into the cosmological processes of change. Numerous images from nature are used to describe self-cultivation such as planting and nourishing seeds, pulling up weeds, refraining from overgrazing land, or avoiding cutting down trees wantonly. Human beings nurture the seeds of virtue within themselves and participate in both the natural and human orders.[12] This is elaborated by the Neo-Confucians through a specific understanding of a correspondence between virtues practiced by humans as having their natural counterpart in cosmic processes. In his "Treatise on Humaneness" Chu Hsi speaks of four moral qualities of the heart-mind of Heaven and Earth: origination, flourish, advantage, and firmness. Similarly in the heart-mind of humans there are four corresponding moral qualities: humaneness, righteousness, propriety, and wisdom. Cosmological powers and human virtues are seen as two aspects of one dynamic process of transformation in the universe.

This anthropocosmic worldview originates in classical Confucianism and conceives the human heart-mind as completing the triad with Heaven and Earth. Through self-cultivation humans affect the growth and transformation of things and create the possibility for a flourishing, sustainable society. The interrelationship of Heaven, Earth, and human is expressed as a parental metaphor, with humans as children of the universe and having a filial responsibility for its care and continuation. The beginning of the *Western Inscription* of Chang Tsai (1020–1077) offers one of the richest articulations of this metaphor:

> Heaven is my father and Earth is my mother, and even such a small creature as I find an intimate place in their midst.
>
> Therefore, that which fills the universe I regard as my body and that which directs the universe I consider as my nature.
>
> All people are my brothers and sisters, and all things are my companions.[13]

Wang Yangming: A Brief Biography

Wang Yangming, the leading Ming Neo-Confucian scholar, statesman, and military leader, was born into a period of enormous upheaval.[14] Wang struggled to overcome great personal difficulties to formulate a system of thought and action that challenged the orthodox Neo-Confucianism of his day. Against formidable odds of political corruption, social decay, and educational mediocrity, he offered fresh and invigorating insights into perennial Confucian problems. While these insights were born out of the intense political and social challenges of Ming China, his hard-won reflections on epistemology, ethics, and cosmology have transcended the particular historical context in which they arose. Indeed, they have been a source of inspiration across East Asia for some five hundred years. As a comprehensive vision of the nature of humans and our capacity for reciprocity with nature, his philosophy bears reexamination in our own times.

Wang was born in 1472 southeast of Hangchow in Yueh in modern-day Chekiang. His birth name was *Shou-jen* (holding on to humaneness). From ages ten to fifteen he lived with his father who was a government official in Peking. It was during this period that his mother died. He was married at age sixteen and on his way home with his wife the following year he met a prominent Neo-Confucian scholar, Lou Liang (1422–1491). Lou urged him to study the thought of Chu Hsi, which he did in earnest after passing his native provincial examination. However, he became disillusioned with these studies when he fell ill trying to penetrate the principle of things, as Chu has urged. He turned to writing flowery compositions and twice failed the national civil service exams. He also had heroic ambitions of suppressing rebels on the borders and he thus studied military tactics and archery.

On his third attempt, when he was twenty-eight, he passed the civil service exam. He thus received a government appointment, first in the department of public works and the next year in the justice department. He became known for his brilliance, but this work took a toll on his health and he became disillusioned with the intellectual life in the capital. He retreated from the capital and built a small house in the mountains in Yangming-tung where he could recuperate and seek spiritual solace and direction. This is where he took his honorific name, *Yangming-Tzu*, philosopher of Yangming. He pursued studies in Taoist longevity cults and Buddhist meditative practices but he found these lacking.

When he recovered his health he assumed government positions once again. He began also to gain followers as he lectured on the true nature of Confucianism to pursue sagehood. He met a renowned scholar, Chan Jo-shui (1466–1560), who encouraged his intellectual and spiritual development. Wang, however, had a drastic turn of fortune when in 1506 he protested the unjust treatment of several officials at the hand of the court eunuch, Liu Chin. He incurred Liu Chin's wrath and was subject to a traumatic public lashing before the emperor. He was then banished to live among the Miao tribal people where living conditions were rugged. Nonetheless, amidst this hardship, he was able to pursue his own understanding of the extension of knowledge and the realization of the unity of knowledge and action. These pursuits came together in an enlightenment experience in 1508.

In 1510 he was restored to public office and received numerous positions over the next six years. During this time his reputation as a teacher spread. He was appointed in 1516 as governor of the area bordering Kwantung, Kiangsi, and Fukien. It was during the next four years that he led several successful military campaigns to overcome rebels. In his efforts to restore order and rehabilitate the rebels, he built schools and promoted education. He also instituted the community compact so that local people could organize themselves for the common good of their region.

In 1519 Wang subdued a rebellion in Kiangsi by Prince Ning, a nephew of the emperor. Ning had declared himself head of state and, backed by a huge army, intended to conquer the capital city of Nanking. Wang was able to capture the prince after ten days of fighting and thus ensure the survival of the Ming dynasty. However, despite his heroic campaign, Wang suffered unjustly at the hands of the emperor who had wished to subdue the Prince himself and claim victory. Wang lost favor until 1521 when the next emperor ascended to power.

In 1521 at age forty-nine, Wang developed his mature doctrine of the extension of innate knowledge. For the next six years he lived in virtual retirement in his native area. During this time his original ideas continued to attract large numbers of devoted students as well as hostile critics. When he lectured, hundreds of scholars came to hear him. His final act of public service came in 1528 when he pacified another rebellion, this time in Kwangsi. While returning from this campaign early in 1529, he died.

Grounds for an Ecological Philosophy and Ethics

Wang's active life, with numerous challenges to overcome, clearly shaped his philosophical ideas and ethical perspectives. He embodied many of the ideals of the Confucian scholar-official who practiced self-cultivation so as to better serve the wider society and bring order to the state. Yet as Tu Weiming observes, Wang transcended the model of a conventional Confucian literati, for his philosophy was born out of immense personal suffering. He described this struggle, often in isolated or difficult circumstances, as "a hundred deaths and a thousand sufferings."[15] Despite these difficulties and in the face of severe criticism, he was able to breathe new life into Neo-Confucianism and to inspire debates on the nature of learning, cultivation, and action. His reflections on epistemology, ethics, and cosmology have significant implications in our current search to formulate sustainable human-Earth relations.

The three key aspects of Wang's thought are his ideas of innate good knowing *(liang chih)*, the unity of knowledge and action *(chih-hsing ho-i)*, and forming one body with all things *(wan wu i-t'i)*. The first outlines an epistemology of an empathy of knowing, the second articulates an ethics of embodied action, and the third embraces a cosmology of kinship of being. Through these ideas Wang articulates a profound sense of reciprocity with all of life—touching the depths of things. Moreover, he understands the human as having a heart-mind that embraces the interconnected circles of society, nature, and cosmos. These ideas lend themselves to ecological philosophy and ethics for they underscore the importance of subjectivity, embodiment, and reciprocity.

In this regard, it might be noted that, although the causes of the current global crisis are manifold—economic, political, social, and technological—a major contributing factor is the imbalance of human relations with the natural world. Because these relations are out of balance, we have created societal norms that are threatening the variety and complexity of life forms on the planet. Our attitudes and our actions regarding nature reflect a dominance model rather than a reciprocity model. Because we have lost or obfuscated the deep connection between humans and the natural world, we have forgotten appropriate boundaries for creating the conditions for flourishing societies. What is needed is the recovery of modes of reciprocal relationship with nature and the cosmos. It is here that Wang Yangming's Neo-Confucian thought may be instructive. By examining some of his key ideas we can reimagine what recovering the depth of things would look like, not only in the Chinese context, but in our own Western milieu as well.

Innate Knowledge of the Good: An Empathy of Knowing

Wang's emphasis on innate knowledge of the good was a central focus of his thought. In highlighting empathy of knowing, Wang was attempting to overcome the fragmentation of learning and the objectification of things that characterized the degenerate forms of Confucianism in his own time. Instead, he wished to affirm the deep wellsprings of human subjectivity as a source of authentic understanding. In so doing, he sought to express his own personal experience of realizing an integrated heart-mind.

He celebrated the primordial numinous quality of the heart-mind, saying, "It is my nature endowed by Heaven, the original substance of my mind, naturally intelligent, shining, clear, and understanding." The heavenly nature of the heart-mind for Wang was identical with the Principle of Nature and with the Tao. It was self-sufficient yet pervasive. He says, "Innate knowledge is man's root which is intelligent and is grown by nature. It naturally grows and grows without cease." He compares it to the "spirit of Creation."[16]

This pure and potential responsiveness of the heart-mind continually preoccupied Wang. He observes, "The mind of the sage is like a clear mirror. Since it is all clarity, it responds to all stimuli as they come and reflects everything" (Wang, 27). Key to Wang's ideas is the intuitive capacity of the heart-mind for sympathetic resonance with all things. He notes that the heart-mind's original substance is sincerity and commiseration (Wang, 176). In its primal form, before the feelings are aroused, it is in a state of equilibrium and thus impartial.

Wang Yangming believed this broad, empathetic basis of apprehension needed to be recovered before things could be properly investigated. Learning should cultivate this innate knowledge so that one could sense the inner depths of things and thus investigate things more clearly. He did not deny the importance of objective knowledge, but felt that the relationship of person and things should not be lost in abstract or disembodied knowledge.

This intuitive awareness was grounded in Wang's personal experience. His life-long struggle to discover and to nurture empathetic knowing is exemplified in his search for principle in things in the world and in texts. A breakthrough moment occurred for him when, after exhausting meditative efforts to understand the essence or pattern of bamboo by looking at it, he finally had an experience of seeing into the true nature of the bamboo. He recognized that seeking principle externally was inadequate for principle lay within the heart-mind itself. This intuitive knowing is translated by Tu Weiming as a "primordial awareness."[17] Indeed, for Wang the cognitive, affective, and ethical intelligence of the heart-mind was one.

The key for uncovering this innate knowledge of the good was making the will sincere so as to eliminate selfish human desires. Purifying one's intentions, attitudes, and motivations was essential for activating this depth of feeling-awareness. Wang Yangming writes, "If the will is sincere, then, to a large extent, the mind is naturally rectified, and the personal life is also naturally cultivated" (Wang, 55). In reference to the *Great Learning* (attributed to Confucius, one of four books promulgated by Chu Hsi as classic Confucian texts), Wang notes, "The task of the *Great Learning* consists in manifesting the clear character. To manifest the clear character is none other than to make the will sincere, and the task of making the will sincere is none other than the investigation of things and the extension of knowledge" (Wang, 86).

Disagreement with Chu Hsi

Wang disagreed with Chu Hsi about the priority of sincerity of the will over the investigation of things. They interpreted the *Great Learning* differently: Wang believed Chu had rearranged the order of the sentences. Although Wang hesitated to challenge Chu Hsi, Wang nonetheless thought it was essential to emphasize learning as based on a sincere attitude. Chu was interested in the investigation of things *(ko-wu)*, while Wang insisted this should be understood as the rectification of things. For Chu this implied objective study; for Wang it implied the search for truth within. Their differences were a matter of emphasis. For Wang genuine learning was not simply investigating facts but understanding the deep interconnected nature of reality—touching the depths of things. This required overcoming the obstacles of selfish desires and distractions.

While Wang stressed the unity of innate knowledge, Chu felt it important to make distinctions. Chu said that moral awareness developed through the study and investigation of principles outside the mind. Wang stressed the primacy of the heart-mind as containing all principles. Chu Hsi believed learning should proceed empirically and deductively by searching unceasingly for principle externally in texts and in things. Wang believed this was not the right procedure; it could become forced or strained. Rather, one needed to recognize that principles reside within. A prior existential determination, illuminating the heart-mind and establishing the will, could not be reached empirically, but could be reached intuitively and inductively. Sincerity was essential to this task.

The Unity of Knowledge and Action: Encouraging Spontaneity of Action

For Wang this cultivation of empathetic knowing grounded in sincerity led to an ethics that emphasized the unity of knowledge and action. When innate knowledge comes into consciousness then spontaneous action is possible. Throughout his life of intense commitments as scholar, statesman, and military leader, Wang demonstrated this fundamental unity of knowing and doing. For him they were part of one spontaneous movement where the mind and body could act in perfect harmony for a larger moral good. He writes: "Knowledge in its genuine and earnest aspect is action, and action in its intelligent and discriminating aspect is knowledge. At bottom the task of knowledge and action cannot be separated" (Wang, 93).

The critique of this profound subjectivity of knowledge is that it could slip into subjectivism. Evil arises due to personal limitations, inertia, and self-deception. Thus sincerity, transparency, and authenticity must continually be cultivated. Safeguards are also required to avoid solipsistic subjectivity. This was the danger that faced many of Wang's followers in the later T'ai-Chou school.

To avoid narcissistic subjectivity, Wang aimed at nourishing innate knowledge in relation to other Confucian scholars, to the broader Confucian tradition, to the larger social community, and to Heavenly principle. He hoped to foster a gradual process of transformation that was effortless *(tzu jan)*. Wang thus aimed to promote embodied action in the world following naturally from one's deep inner springs of innate wisdom. He aspired to spontaneity that would bring the individual into a vibrant relationship with other humans and with all forms of life. From such a balanced inner unity, harmonious action would result. Humans would be reciprocal with other humans and resonate with the natural world at large. He expresses this spontaneity in a poem written in 1524 after he had retired to his native place:

Sitting at Night at Pi-Hsia Pond

An autumn rain brings in the newness of a cool night:
Sitting on the pond's edge I find my spirit brightened
by the solitary moon.
Swimming in the depths, the fish are passing on words of power;
Perched on the branches, birds are uttering the true Tao.
Do not say that instinctive desires are not mysteries of Heaven:
I know that my body is one with the ten thousand things.
People talk endlessly about rites and music;
But who will sweep away the heaps of dust from the blue sky?[18]

Forming One Body With All Things: Kinship of Being

Thus from innate knowing and the unity of knowledge and action comes Wang's rich cosmological understanding of forming one body with all things. This microcosmic-macrocosmic identity characterizes Wang's thought and draws on many strands of earlier Confucian ideas. From classical notions of forming one body with the myriad things in such canonical texts as the *Book of History* and the *Doctrine of the Mean* to Chang Tsai's filial relationship with Heaven and Earth in the *Western Inscription*, Wang's organismic identity broadens and deepens the Confucian vision. His resonant language for expressing this identity embraces not only the human order but the natural order as well. What distinguishes his thought is how he extends sympathy and concern beyond humans to animals, birds, plants and living things, and finally even to inanimate objects. The ecological implications are manifold in the following selections from his writings:

"That the great man can regard Heaven, Earth, and the myriad things as one body is not because he deliberately wants to do so, but because it is natural to the humane nature of his mind that he do so. . . . When he observes the pitiful cries and frightened appearance of bird and animals about to be slaughtered he cannot help feeling an 'inability to bear' their suffering. This shows that his humanity forms one body with birds and animals. . . . Even when he sees tiles and stones shattered and crushed, he cannot help a feeling of regret. This shows that his humanity forms one body with tiles and stones." (Wang, 272–3).

This anthropocosmic cosmology highlights the identification of the microcosm of the human with the macrocosm of the Earth and the universe. The interconnected concentric circles that move outward from humans to the larger community of life is evident. That these circles extend even to inanimate things indicates the depths of our relationality.

The Ecological Implications of Wang's Thought

Wang Yangming encouraged a profound affective identification of humans with the natural world. He recognized the nature of the heart-mind as having the capacity for sympathetic resonance with all things, mirroring and responding to the life forces around it. Wang provided an interrelated means of self-cultivation beginning with an epistemology that nourished empathetic knowing, moving into an ethics encompassing the unity of knowledge and action, and ultimately entering into the cosmological interpenetration of self, society, nature, and cosmos.

For Wang, the principles embedded in the heart-mind are the means of understanding the larger world. In order to examine principles within, one has to make the will sincere. Then the extension of knowledge is possible. Ultimately through sincerity one can maintain the mirrorlike quality of the heart-mind. One can thus reflect the principles of nature, extend one's affective knowing, and link to the sincerity of the larger universe as suggested in the *Doctrine of the Mean.*

Wang's teachings on innate knowledge are important as a means of reawakening the capacity of humans to apprehend nature and embrace its vast complexity and magnificent particularity. An appreciation of the principles of nature's diversity and patterns is indispensable in overcoming current tendencies to objectify nature as external to humans and thus to use it mindlessly.

This deepening knowledge of nature encourages reciprocity rather than dominion, communion rather than exploitation. Wang's life-long aspiration to touch the depths of things echoes the yearnings of our present moment for humans to embrace their larger self: "Everything from ruler, minister, husband, wife, and friends to mountains, rivers, spiritual beings, birds, animals, and plants should be truly loved in order to realize my humanity that forms one body with them and then my clear character will be completely manifested, and I will really form one body with Heaven, Earth, and the myriad things" (Wang, 273). Our challenge is how to realize our humanity within the larger Earth community in this new millennium.

NOTES

1. Joy A. Palmer, ed., *Fifty Key Thinkers on the Environment* (London: Routledge, 2001).
2. For a broad survey of some of the key thinkers in the Confucian and Neo-Confucian tradition, see Tu Weiming and Mary Evelyn Tucker, ed., *Confucian Spirituality*, 2 vols. (New York: Crossroad, 2003, 2004).
3. This term is used by Tu Weiming in *Centrality and Commonality: An Essay on Confucian Religousness* (Albany: State University of New York Press, 1989).
4. This section on naturalistic cosmology and the following on transformative ethics are adapted from my introduction to *Confucianism and Ecology: The Interrelationship of Heaven, Earth, and Humans*, Mary Evelyn Tucker and John Berthrong, ed. (Cambridge: Harvard Divinity School Center for the Study of World Religions, 1998), xxxvi–xxxviii.
5. See Cheng Chung-ying, "The Trinity of Cosmology, Ecology, and Ethics in the Confucian Personhood" in *Confucianism and Ecology*, Tucker and Berthrong, 212–235; quotes from 228, 215. For more on these concepts, see Cheng Chung-ying, *New Dimension of Confucian and Neo-Confucian Philosophy* (Albany: State University of New York Press, 1991), particularly 4.
6. Frederick F. Mote, *Intellectual Foundations of China* (New York: Knopf, 1971), 17–18.
7. See Chapter Two,"The Continuity of Being: Chinese Visions of Nature," in Tu Weiming's *Confucian Thought: Selfhood as Creative Transformation* (Albany: State University of New York Press, 1985).
8. Tu, *Confucian Thought*, 39. Professor Tu notes, "For this reference in the *Chou I*, see *A Concordance to Yi-Ching*, Harvard Yenching Institute Sinological Index Series Supplement no. 10 (reprint; Taipei: Chinese Materials and Research Aids Service Center, Inc., 1966), 1/1."
9. See how this developed in Japan in Mary Evelyn Tucker, *Moral and Spiritual Cultivation in Japanese Neo-Confucianism* (Albany: State University of New York Press, 1989).
10. For discussion of debates on *li* and *ch'i* in China and Japan see Kaibara Ekken, *The Philosophy of Qi*, ed. and trans. Mary Evelyn Tucker (New York: Columbia University Press, 2007).
11. See Tu, *Confucian Thought: Selfhood as Creative Transformation*; even its subtitle contains this idea.
12. See Sarah Allan, *The Way of Water and Sprouts of Virtue* (Albany: State University of New York Press, 1997).
13. Chang Tsai, "Western Inscription," in *A Source Book of Chinese Philosophy*, trans. and ed. Wing-tsit Chan (Princeton: Princeton University Press, 1963), 497.
14. For biographies of Wang Yangming see Tu Weiming, *Neo-Confucian Thought in Action: Wang Yang-ming's Youth (1472–1509)* (Berkeley: University of California Press, 1976) and Julia Ching, *To Acquire Wisdom: The Way of Wang Yang-Ming* (New York: Columbia University Press, 1976).

15. Tu, *Neo-Confucian Thought in Action*, 4–5.
16. Wang Yangming, *Instructions for Practical Living and Other Neo-Confucian Writings*, trans. Wing-tsit Chan (New York: Columbia University Press, 1963), quotations from 278, 210, 216 respectively, hereafer page numbers are cited in the text.
17. Tu, *Confucian Thought*, 32.
18. Wang Yangming, "Sitting at Night at Pi-Hsia Pond," transl. Julia Ching in Ching, *To Acquire Wisdom*, 237. Poem reprinted by permission of John Ching.

Interiority Regained: Integral Ecology and Environmental Ethics

Michael E. Zimmerman

Having wrapped themselves for decades in the mantle of the natural sciences in order to gain credibility, some environmentalists concede that this strategy involves a serious problem. While valuing the natural world highly and calling on others to join them in protecting it, environmentalists typically subscribe to a naturalistic materialism that either excludes values altogether, or else regards them socio-biologically as an adaptive strategy useful for enhancing the survival chances for a certain species. For such naturalism, all phenomena are assumed to have only *exterior* aspects that can be analyzed wholly in third-person terms, that is, as "its." In a world comprised of *its*, there is no place for you, me, and us, that is, for the first- and second-person *interior* domains that comprise aesthetic experience, morality, consciousness, subjectivity, freedom, values, intersubjectivity, and culture, *considered in their own terms*. Naturalism maintains that for something to be, means for it to be a phenomenon analyzable without remainder either into its externally observable parts and behaviors, or else into its functions within an externally observable system. From the perspective according to which there are only *its*, including human *its*, all talk of interiority is naïve, retrograde, and misguided.

Many environmentalists conclude that contemporary environmental problems often arise from anthropocentrism, according to which human rationality, consciousness, subjectivity, soul, or moral capacity justify dominating nonhuman beings. For such environmentalists, naturalism has worked against anthropocentrism not only by denying that humans are the culmination of evolutionary purpose, because there is no such purpose, but also by eliminating the entire domain of interiority with which spirituality, subjectivity, consciousness, values, and culture are associated. By pulling the rug out from under human arrogance, environmentalists hope to save natural habitat from destruction. Ending cosmic hierarchy displaces humankind from its throne atop the tree of life, for there is no such tree. Instead, life is a bush with many branches, no one of which is higher or more central than any other.

In support of what some deep ecologists called radical biocentric egalitarianism, other anti-hierarchical environmentalists adopt the language of systems theory, according to which all phenomena are merely functional strands in the great cosmic web. In this web, humans have no special standing *vis-à-vis* plants, animals, and even physical habitat. Instead, humans are merely one animal species among others striving to enhance their reproductive advantage over competitors. Read in socio-biological terms, concepts such as rights and values are merely useful fictions that bestow a measure of justification to nature-domineering practices aimed at enhancing the power of humankind over other life forms. In view of all this and more—much more—environmental ethics not only seems impossible, but worse still: it is an embarrassment to the scientifically sophisticated.

Neither biologist nor naturalistic environmentalist would offer a *moral* critique of *non*human species that maximize their reproduction. Population crash and even extinction may result if a species were to overshoot the carrying capacity of its habitat, but there would be no moral failing involved.[1] For the human animal, too, only a *prudential* "ought" can be recommended: We ought to alter our behavior toward the nonhuman domains to promote long-term human survival. Yet, many environmentalists insist that a *moral* "ought" also applies here: We ought to limit our behavior, including our reproductive drive, so that *other* life forms can survive and prosper. The (often tacit) presupposition that only humans are morally responsible for their behavior conflicts with naturalism, which has difficulty accounting for morality, values, and "oughts" *in their own terms*, along with other contents of *interior* domains to which access cannot be gained by the third-person methods of the natural and social sciences.

Integral Ecology: Theoretical Considerations[2]

For those unsatisfied with the paradoxical situation in which environmental ethics is undermined by the naturalism favored by many environmentalists, there are some alternatives. One is the deservedly influential work of the philosopher, Holmes Rolston III, who argues that human beings have a capacity for discerning extra-human moral and aesthetic value, just as they have the capacity for discerning galaxies or amino acid compounds. Another alternative, which I explore in this essay, is Ken Wilber's integral theory, according to which the cosmos is an infinite set of nested wholes, the constituents of which can be described only in terms of multiple perspectives, including those involving the third person (the natural and social sciences), the second person (the humanities), and the first person (phenomenology and fine arts). Wilber insists that however successful the methods of the natural and the social sciences may be, those particular methods pertain only to third-person phenomena, that is, to objects. Without interiority, objects lack the capacity—however meager it may be—to constitute a perspective of their own in which to register or to take account of other things. By restoring an *appropriate* position for interiority in the cosmos, we can solve two related problems.

The first is the widespread modern sense that self-aware humankind is an accident in a meaningless material universe within which there is no place for awareness of any sort. Alienated from body, emotions, nature, and even consciousness, the abstract modern human ego sets out to know and control the material world that seems simultaneously very real and yet completely other. The second problem is the widespread eco-romantic conviction that only by renouncing mind, consciousness, rationality, and other allegedly alienating features of human consciousness, and only by reabsorbing themselves within the patterns of nature, that is, only by scraping off the sorry accretions imposed by 10,000 years of civilization, can human beings regain their lost unity with Mother Nature. Despite important differences, both the modern abstract ego and the eco-romantic self share something in common: an industrial or flatland ontology, according to which the only things that exist are the surfaces or exteriors of things as they appear in terms of third-person perspectives.

Wilber, however, maintains that *interiority* is a basic feature of reality, every bit as real in its own way as mass, energy, space, or time. Moreover, interiority is not restricted to humans or even to animals; instead, interiority goes *all the way down*. According to those who work in the field of bio-semiotics, signaling is a universal feature of life and perhaps of nonlife as well. Signaling cannot be understood merely externally as some mechanical interaction, but points to interior domains—first- and second-person perspectives, however constricted these may be—that correspond to the externally visible aspect of the signaling that scientists study from the third-person standpoint. What would happen to our conception of terrestrial nature at many different scales if researchers assumed that the phenomena they study—such as animals and plants—had an interior aspect that would have to be taken into account in order fully to characterize the phenomenon in question?[3] How would the science(s) of ecology have to change? Indeed, how would *environmentalism* have to change, perhaps especially with regard to its troubled relationship with human beings, who are endowed with such an extensive, linguistically enhanced mode of interiority?

Most environmentalists abjure talk of transcendence and spirit because they are moderns at heart, that is, they agree that all being is material being. Moreover, because mainstream religions have traditionally limited transcendence (along with interiority or "soul") to humans, environmentalists fear that transcendence-talk only encourages a version of the anthropocentrism that justifies heedless human destruction of natural phenomena. Eco-romantics "think transcendence is destroying Gaia, whereas transcendence is the only way fragments can be joined and integrated and thereby saved."[4] Genuine transcendence is neither anthropocentric nor otherworldly, but always integrated with the world in the nondual embrace described in Mahayana Buddhism's dictum that *samsara* (the cycle of birth and rebirth) is not other than Nirvana.

Drawing on Wilber's work, integral ecologists envision a postindustrial ontology that restores depth to the cosmos by reintegrating what has been dissociated, i.e., the interior, subjective domains that characterize all phenomena. Following Max Weber and Jürgen Habermas, Wilber notes that modernity differentiates among three domains, which he calls the Big Three: 1) consciousness, subjectivity, self, and self-expression (including art), whose mode of truth involves truthfulness and sincerity; 2) ethics, morality, worldview, culture, intersubjective meaning, whose mode of truth involves justice; 3) natural and social science, whose mode of truth involves correct propositions.[5] These three differentiated domains generated the social-cognitive space necessary for free scientific inquiry, new art forms, market economies, and democratic politics, including liberation movements ranging from abolitionism to feminism.

Unfortunately, modernity did not adequately integrate the three domains. Because personal-artistic and cultural-moral truth claims are more complex and contentious than those made by empirical scientific research, and because scientific knowledge brought such important material gains, scientific modes of knowledge marginalized the other two. Natural science could not even notice, much less study, selfhood, interiority, culture, and morality, since empirical inquiry is suited for material phenomena, not for personal and social phenomena. Far from representing nature as a sum of disconnected atoms, as some environmentalists have complained, modern science represented nature as "a perfectly *harmonious and interrelated system*, a great-it-system, and knowledge consisted in patiently and empirically mapping this it-system." Modern science unified the cosmos in terms of the "great 'web of life' conception, a great interlocking order of beings, each mutually interwoven with all others."[6]

The modern rational ego sought to disenchant nature, both to eliminate any lingering concerns about violating "Mother Nature" and to achieve the ideal of rational and moral objectivity. So long as one's reasoning processes are influenced by biological factors (e.g., emotions), so long as one's moral judgment is tainted by personal, familial, tribal, or racial factors, one is not truly rational, impartial, and thus fully human. Following Kant, the modern ego sought to overcome the domain of particularity and corporeality, in order to attain universality and impartiality. But this quest had two major problems. First, moderns could not really admit to a domain transcending the material plane; hence, the ego was left in a transcendental limbo that was made increasingly untenable by the relentlessly reductive processes of scientific materialism. To make up for its own conceptual erasure, so Wilber argues, the modern ego engages in extraordinary, nature-dominating *agency*. To demonstrate its own existence, in other words, the ego set out to subjugate the material domain, i.e., the only domain that supposedly exists. Martin Heidegger, as well as Max Horkheimer and Theodor Adorno, claimed that the striving of modern "man" for world domination showed that he had become an animal seeking power and security. Wilber holds a quite different view. The striving for world domination represents, at least in part, an effort at self-assertion on the part of persons who *intuit* their own (interior and interpersonal) reality, but who cannot find any adequate personal or cultural expression for it. Hence, when Marx said that the point of philosophy is not to reflect on the world, but rather to change it, he sought in part to reemphasize the power of human agency in a world that was increasingly mechanized and devoid of subjectivity.

The second problem with the modern quest for universality was that the justifiable differentiation between mind and body ended up in unjustifiable dissociation:

> The rational ego wanted to rise above nature and its own bodily impulses, so as to achieve a more universal

> compassion found nowhere in nature, but it often simply repressed these natural impulses instead: *repressed* its own biosphere; repressed its own life juices; repressed its own vital roots. The Ego tended to repress both external nature and its own internal nature (the id). And this repression, no doubt, would have something to do with the emergence of a Sigmund Freud, sent exactly at this time (and never before this time) to doctor the dissociations of modernity.[7]

The romantic reaction against rational modernity's humanity-nature split, and against the repression that follows from it, was justified, for something serious was amiss. Nevertheless, romantic efforts to heal this rift went astray because they employed two competing conceptions of nature. The first was the modernist view that nature is the all-encompassing, interrelated web-of-life. Supposedly, modernity had lost touch with this web-of-life, despite asserting that everything is enclosed and flows within it. In positing that culture has deviated from or split off from nature, then, the romantics had to posit a *second* conception of nature, a nature from which humankind *can* deviate. Wilber asks: "What is the relation of this Nature with a capital N that embraces *everything*, versus this nature that is *different* from culture because it is getting ruined by culture?" Romanticism foundered because it could not reconcile these conflicting views of nature. Great romantics, such as Schelling, sought to reconcile this conflict by identifying Nature with an all-embracing Spirit that transcends and includes both culture and nature. Most romantics, however, "simply identified Nature with nature."[8]

Arguably, back-to-nature fantasies reprise this failed romantic effort to overcome the humanity-nature split. "Instead of moving *forward* in *evolution* to the emergence of a Nature or Spirit (or World Soul) that would indeed unify the differentiated mind and nature, [romantics] simply recommend 'back to nature'." The quest for this kind of unity invites psychological and social regression. If nature as biosphere is the "fundamental reality" (Goddess/Gaia), then that which deviates from nature threatens nature. If nature "is the ultimately Real, then culture must be the original Crime." The goal, then, must be to dismantle culture in order to achieve a lost paradise involving unconscious unity with pristine nature. Such a yearning for primal unity with divine nature is tempting, but potentially disastrous both individually and culturally. Moreover, eco-sentimentalism will not halt ecological destruction. Required is a major change in socioeconomic and political institutions, but such change occurs only in tandem with the interior growth and development consistent with "mutual understanding and mutual agreement based upon a worldcentric moral perspective concerning the global commons."[9]

Although criticizing anthropocentrism for wrongly conceiving of subjectivity or consciousness as an exclusive human property, integral ecology also affirms that humans are endowed with a distinctively rich, linguistically articulated mode of interiority. This same interiority makes possible not only the technological power to exploit nonhuman beings, but also the possibility of developing to a moral level capable of calling for limits to human action out of respect for nonhuman beings. What is often called the technological domination of nature could have arisen only in the modern era, with its extraordinary combination of natural science and dynamic economic systems. To significantly limit or transform this dominator hierarchy, in which technologically outfitted humans heedlessly exploit nonhuman nature, more is required than changing the social system or developing new technologies. Corresponding to any exterior sociopolitical, economic, or technological dominator hierarchy is an interior dominator hierarchy. Although modern science and technology have often been used by one group of humans to dominate other groups, the same development of consciousness that generated science and technology was also at work generating worldcentric ethical positions, which affirmed the universal rights of humankind. Environmentalists have explicitly called upon such rights to life and property in working against the untoward consequences of industrial technology, such as water, ground, and air pollution. An interior dominator hierarchy remains in place for most modern people with regard to nonhuman beings. Until a critical mass of people move to postmodern levels of interiority, in which heedless domination of nonhuman beings becomes unacceptable and immoral, environmentalism will remain a reform movement within technological modernity. When this developmental move occurs, but not before, we will see the widespread adoption of what Hans Jonas called "the imperative of responsibility" for the kind of world we want to leave to our human descendants, as well as to those of other species.[10]

This brings us to another major feature of integral ecology—its emphasis on the development that occurs in both exterior and interior aspects of phenomena. Using a term coined by Arthur Koestler, Wilber claims that the basic constituents of the cosmos are holons, which range from atoms through organisms. Holons are simultaneously parts of larger wholes and include within themselves as parts holons that are less complex. Cells contain and are thus senior to molecules, while cells are contained in organisms and thus are junior to them. In addition to individual holons, there are social holons. The latter differ from the former for several reasons, including the fact that social holons lack the relatively centered kind of interiority that belongs to individual holons.

Every individual holon may be understood as containing four major features, which can be exemplified by the individual human being:

- The human has an individual exterior that includes the body and all its organs, as well as all the behaviors observable from the third-person perspective;
- The human being is also a member of highly complex social structures, which can also be studied by social sciences from the third-person perspective;
- The human being is also endowed with interiority, hence, constitutes a first-person perspective that correlates with neurological events and with externally observable behaviors;
- Finally, the human being is a member of a culture (or cultures) that arises in connection with mutual exchanges of recognition and communication between first and second persons.

The interiors of cultures, including values, religious belief, philosophy, shared purpose, and so on, correlate with various social structures, ranging from political institutions to economic activity. Wilber uses the term "quadrants" to refer to these four basic constituents of holons.

Every holon contains these four different dimensions that correlate with the four major perspectives people can use to analyze any complex phenomenon: first person, second person, third-person individual, and third-person collective. For example, if I wish fully to understand a small mammal, I must attempt to understand not only its behavior (third-person, individual, objective perspective) and its membership in its extended kin base and ecosystem (third-person, collective, interobjective perspective), but also what it is like to be that mammal (first-person perspective) and what is involved in the mammal's intersubjective domain, including its kin base (second-person perspective). As a shorthand measure, Wilber reduces the four perspectives to what he calls The Big Three (see figure 1): I, You/We, and It(s). Hence, integral ecologists state there are three basic perspectives—first person, second person, and third person—in terms of which to analyze any holon, because holons are such that all three perspectives must be brought into play to fully describe holons as constituents of the cosmos.

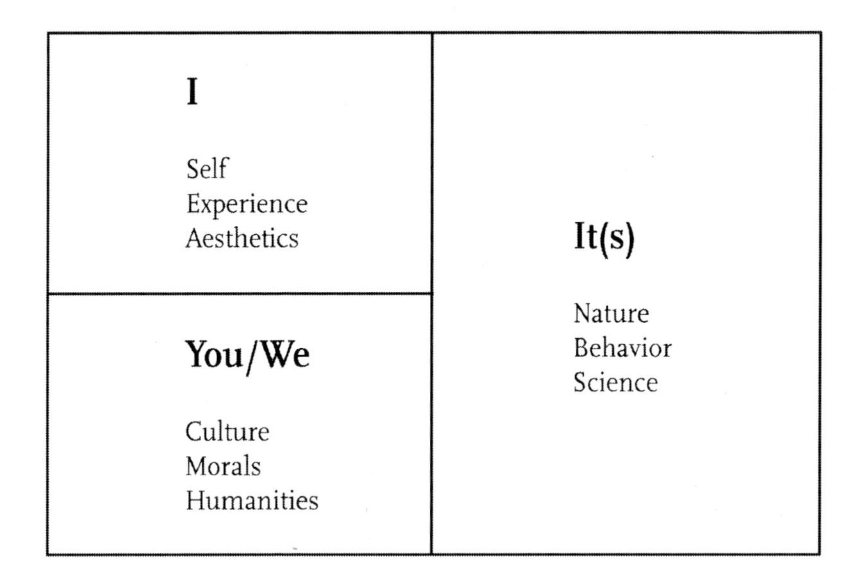

Figure One:
The Big Three

The developmental processes that have characterized history in all quadrants have opened up an indescribably vast hierarchy of different individual and intersubjective perspectives. The perspective taken by a cell establishes a horizon within which phenomena pertinent to the cell can show up. Such a perspective is enormously greater than that of an atom or an inorganic molecular complex, such as a crystal. Likewise, the perspective opened up by a human being is much greater than the perspective held open by a deer mouse. The human perspective expands as the human being matures. Hence, the first-person perspective of a three-year-old child is very different from the first-person perspective of that same person thirty years later. As individuals develop in various domains—whether cognitive, moral, psychosexual, aesthetic, interpersonal, or spiritual—they become capable of generating judgments that are more inclusive, more comprehensive, and more integral.

An integral ecologist is someone who knows that in order to characterize and to devise a solution to an environmental problem, he or she must not only seek insight from many different perspectives, including *the big three*, but must also take into account the different *developmental levels* of people speaking from those perspectives as well as the developmental levels of the public audience to whom such judgments are addressed. Most environmental problems are complex not only in the sense that they are hard to define from the perspective of a particular branch of ecological science, but also in the sense that they are multifaceted such that many different perspectives—cultural, social, political, legal, ethical, religious, and aesthetic—must be utilized to allow the problems to show themselves adequately.

By restricting the term "ecology" to natural science methodologies, that is, third-person perspectives on the exterior facets of individuals and/or collectives, we overlook the fact that dozens of ecological schools adopt a first- or second-person perspective on environmental issues. Integral ecology provides a comprehensive theoretical framework for classifying and coordinating these manifold perspectives. For humans, of course, final or absolute knowledge is impossible because the whole is infinite and still unfolding, whereas human knowledge is finite and perspectival. Yet I assume that it is possible to generate ever more encompassing and inclusive models of this enormously complex and evolving tapestry. Such models provide people with greater capacity for comprehending and intervening in any kind of complex problem, including environmental ones.

Resisting method hegemony, integral ecology is methodologically pluralistic. Integral Methodological Pluralism (IMP) is a collection of practices and injunctions guided by the intuition that everyone's practices bring forth and disclose a different facet of reality. IMP contains three principles: *nonexclusion* (acceptance of truth claims that pass the validity tests for their own paradigms in their respective fields); *enfoldment* (some sets of practices are more inclusive, holistic, comprehensive than others); and *enactment* (phenomena disclosed by various types of inquiry will be different depending in large part on the quadrants, levels, lines, states, types, and bodies of the researchers used to access the phenomena). Wilber describes this commitment to a transmethodological or integral approach:

> The whole point about any truly integral approach is that it touches bases with as many important areas of research as possible before returning very quickly to the specific issues and applications of a given practice. . . . An integral approach [. . .] is a panoramic look at the modes of inquiry (or the tools of knowledge acquisition) that human beings use, and have used, for decades and sometimes centuries. . . . All of the numerous practices or paradigms of human inquiry—including physics, chemistry, hermeneutics, collaborative inquiry, meditation, neuroscience,

> vision quest, phenomenology, structuralism, subtle energy research, systems theory, shamanic voyaging, chaos theory, developmental psychology—all of those modes of inquiry have an important piece of the overall puzzle . . . .[11]

Integral ecology also reinstates the reality and importance of holarchy: some truth claims are better—more inclusive, more comprehensive, more insightful, more generative—than others. In the case of the natural sciences, integral ecology affirms that truth claims arising in this domain must be taken very seriously when it comes to describing environmental problems. Natural science is not a kind of poetry because science and poetry involve very different methodologies. Nevertheless, natural scientific truth claims must not be allowed to trump truth claims generated by competent practitioners in other domains.[12] Objective claims can be contested by other objective claims and subjective claims judged by subjective claims, so within their respective domains, there is better or worse. One ought not, however, judge a subjective claim by an objective standard because the criteria for truth claims are domain dependent.

Being inclusive does not mean abandoning rigor, but it does require that many different kinds of rigorous inquiry be brought to bear on complex problems. Each discipline has its own methods, practices, injunctions, and community standards in connection with knowledge production. People tend to regard their own approach to environmental issues as the only valid one, or at least considerably more valid than the available alternatives. Such an attitude is not restricted to those who deal with environmental problems, of course. Experts in any given discipline tend to regard their particular approach—their method, their perspective, their way of interpreting things—not only as optimal, but also often as exclusively valid. Clearly, this attitude is inconsistent with integral ecology's call for an all-quadrant approach to characterizing and resolving environmental problems. Moreover, an integral ecologist presupposes that any complex phenomenon—such as "wild animal habitat"—will be named, described, and assessed differently by different communities at different points of development.

An integral thinker no longer identifies with the perspective associated with a specific developmental stage, for instance, premodern conservative, or rational modern, or postmodern multicultural, but instead first recognizes how a given phenomenon—such as habitat loss created by logging—shows up differently from within people occupying each of those stages, and secondly recognizes that there are "warrant to truth" claims made within the perspectives constituted by different stages of development.

To ascertain the character and consequences of environmental problems, an integral ecologist must solicit first-person accounts (including testimony, diaries, letters, documentaries, and art works) from people affected by or concerned about environmental problems. Hearing the fear, anger, and suffering of people whose health, families, livelihood, or way of life may be harmed by environmentally destructive practices can have a profound effect on how one evaluates such practices. If the point of view of nonhuman life forms were seriously taken into account as well, extinguishing a species or destroying countless plants and animals would require a higher level of justification than is currently required by most environmental impact statements.

Integral ecology also calls on second-person or cultural perspectives. To garner broad support for an initiative to limit habitat loss, an inclusive rhetorical strategy is needed. People exist within a complex of cultural beliefs, attitudes, practices, norms, and interpretative categories. In recent years, ecofeminists, Third-World representatives, and environmental justice advocates have criticized mainstream environmentalists as well as many environmental philosophers for assuming that the white, middle-class, male, American environmentalist perspective (including beliefs, values, norms) is the *true* way of disclosing humanity's relation to nature, rather than a particular way. Mainstream environmentalists who depict genuine nature as "wilderness," that is, pure land untouched by humans, are influenced by aesthetic, moral, and other cultural categories that are not necessarily shared by others. In characterizing an environmental problem, an integral ecologist recognizes the need to discern and to take seriously competing cultural perspectives on what constitutes "nature," as well as on what constitutes beauty, goodness, justice, fairness, compassion, and so on. Indeed, what manifests itself as a *problem* differs from one culture to another.

Examining social and systemic phenomena from a third-person perspective, social scientists attempt to generate knowledge claims that enable them to make predictions about the publicly observable behavior of social groups. Marx's idea that the economic and technological base determines the cultural and personal superstructure has proven to be a powerful presupposition for social science. Marx's point is that individual behavior is largely a function of social roles determined by socioeconomic factors, which in turn are profoundly influenced by technological innovation. For example, thousands of years ago the introduction of techniques for large-scale agriculture gave rise to urban life, which made possible a host of social roles unavailable to horticultural and gatherer-hunter societies. Just as the invention of steam power paved the way for industrialization and its attendant social upheaval, so too the digital revolution will redefine human possibilities in ways that cannot even be foreseen. Individual behavior may vary, but only within the perimeters laid down by overriding social, political, and economic structures. Just so, many ecosystem biologists maintain that individual organisms are primarily functions of their species, which in turn are shaped in part by the prevailing environmental circumstances. Hence, habitat protection—not protection of individual organisms—is high on the list of many environmental activists, whose views are largely shaped by ecosystem biology. Despite the power of social structural analysis, an integral ecologist insists that such analysis in and of itself cannot provide a complete description of or resolution to environmental problems.

Finally, the natural sciences use powerful third-person methods to generate objective truth claims about a vast array of phenomena, ranging from sub-atomic particles and molecules, to cells and organisms. Although ideally providing the "facts of the matter" to which all parties will agree, the findings of even the "hard" sciences have become subject to contestation by critics who point out the growing dependence of researchers on corporate funding. Moreover, the ideal of investigative objectivity is challenged by factors such as gender, race, class, and social structure, which inevitably influence the process of knowledge formation. The ideal of objectivity without presuppositions is unrealizable for finite beings that must examine things from a particular perspective, using a particular method.

Nevertheless, *some* degree of valid knowledge can be achieved. By examining environmental problems from a host of different perspectives, and by taking seriously the disagreements that arise among people operating from within the same perspective, integral ecologists envision nothing more than attaining the most comprehensive understanding of those problems that finite knowers can achieve at present. They assume that more powerful paradigms will emerge for understanding and acting upon phenomena.

Aldo Leopold's Land Ethic Anticipates Integral Ecology

Only recently did I realize that Leopold anticipated two of the major features of integral ecology in his famous land ethic, published in 1949.[13] First, he affirmed that in addition to natural and social science perspectives, people need to bring to bear ethical, cultural, and aesthetic perspectives on land use (environmental) issues. Using terms drawn from integral ecology, we could say that Leopold distinguished between exterior, third-person perspectives, and interior, first- and second-person perspectives. Second, he argued that only an evolutionary development in human morality would make it possible for those perspectives to be taken seriously at the negotiating table.

> The extension of ethics to this third element [the land] in human environment is, if I read the evidence correctly, an evolutionary possibility and an ecological necessity. . . . Individual thinkers since the days of Ezekiel and Isaiah have asserted that the despoliation of land is not only inexpedient but also wrong. Society, however, has not yet affirmed their belief. I regard the present conservation movement as the embryo of such an affirmation.[14]

Despite the fact that Leopold's account of ethics and ethical evolution is relatively simplistic, he understood that ethical and aesthetic perspectives would not be included in land issue debate until significant ethical evolution took place.

> The 'key-log' which must be moved to release the evolutionary process for an [environmental or land] ethic is simply this: quit thinking about decent land-use as solely an economic problem. Examine each question in terms of what is ethically and esthetically right, as well as what is economically expedient. A thing is right when it tends to preserve the integrity, stability, and beauty of the biotic community. It is wrong when it tends otherwise.[15]

The land is a complex community comprised of the earth's many different habitats and their associated life forms. Leopold knew that the third-person methods used by natural science provide important insights into the land and land-use. Such methods allowed him to develop the idea of the land pyramid, according to which solar energy is first captured by plants, after which it slowly works its way up the food chain to top predators. All the way up the chain, dying organisms return vital nutrients to the land. Leopold also knew how important economics and other social sciences were for informing decisions about land-use. However, he concluded that insights afforded by these third-person perspectives had proven insufficient to prevent short- and long-term damage to the land. In 1949, most people were not prepared to take seriously his summons to consider not only material and economic aspects of the land, but also aesthetic and ethical aspects. Having spent years in the regulatory trenches, having trained in the natural sciences influenced by positivism and behaviorism, he did not underestimate the difficulty of introducing aesthetic and ethical considerations—that is, *interior*, first- and second-person considerations—into decisions that would affect habitat and species. In fact, he postulated that an evolutionary advance is required to move beyond the instrumentalist view, according to which the land is merely raw material for human ends. Given the obstacles facing such an advance, and the time required to achieve it, Leopold knew, too, that he was offering an ecology for the future. Likewise, the conditions for what I am calling integral ecology are not yet in place, although we are moving in the right direction.

A thing is right, Leopold proposed, if it tends to preserve the beauty, integrity, and stability of the land. By "beauty," Leopold evidently had in mind an objective feature of the natural world; by "integrity" he meant the wholeness of the complex fabric of the land, and by "stability" he meant the predictably recurring patterns of the interrelated constituents of the land. Leopold wrote prior to theoretical trends that emphasize intersubjectivity and perspectivalism, and that are skeptical about unchanging foundations. Nevertheless, surely he knew that beauty, integrity, and stability included other and more complex meanings than the ones indicated above. Perhaps he would agree with the following free interpretation of his important claims. Beauty is an assessment of the land, an assessment made by an observer from the first-person perspective, influenced by cultural, socioeconomic, linguistic, and developmental factors. Integrity refers not only to the integrated land-tapestry, but also to the moral rectitude of members of the land community—human beings—who are capable of taking the position of the Other, the second-person perspective. People of integrity can respect the land as Other, and whenever possible can resolve to work with other people to preserve the land's integrity, in the sense of its well-being and wholeness. Finally, for Leopold, stability is an aspect of the land as studied from the third-person perspective of natural science. Stability means not stasis, permanence, or solidity, but instead resilience, the capacity of the dynamic land-community consistently to re-establish its imbricated patterns in the face of perturbations.

As noted earlier, Leopold's land ethic anticipates important features of integral ecology, which is an example of an *ethos*, the Greek root for our term "ethics." *Ethos* means the character, disposition, or fundamental values of particular individuals or communities. Members of the future community of integral ecologists will ideally exhibit respect for and take into account as many perspectives as possible; appreciate and promote the beauty, integrity, and stability of the land; and recognize that the community's truth claims are limited, as well as dependent on the achievements and shortcomings of others.

Leopold reports that his own third-person, objectifying, instrumentalist attitude toward nonhuman life began to change as a young man, when he and some friends were hunting deer. Spotting a pack of wolves, he shot a wolf and one of her cubs, members of a species that was then regarded as a worthless and dangerous predator. As he approached the dying mother wolf, he observed "a fierce green fire dying in her eyes."[16] At that moment, Leopold encountered the wolf as Other, that is, he acknowledged that the wolf had a wolfish kind of first-person sentience and a second-person relation to him. Far from being merely a behavioral mechanism, the wolf exhibited something akin to the yearning, desiring, and fearing that Leopold himself experienced from his own first-person perspective. The wolf had a life of its own, as well as a very important role to play as a member of its ecosystem. To understand the wolf required more than weighing and measuring it, analyzing the working of its organs, studying its behavior, and comprehending its function as a top predator in desert mountain country. Instead, an additional effort was needed to appreciate what it must be like to be a wolf, an otter, a beaver, or a squirrel. Many of the chapters in Leopold's classic, *A Sand County Almanac,* are sympathetic sketches of what it must be like to be an animal trying to make a living in a challenging environment. If behaviorists in the 1940s refused to countenance terms such as subjectivity, consciousness, and awareness in studying human beings, they were even more adamant about denying interiority, inwardness, awareness, subjectivity, and first-hand experience to nonhuman beings.

Leopold's obvious perceptiveness about the lives of animals did not prevent him from being an avid hunter. Like many environmentalists, he assumed that what counts is preserving species of plants and animals, rather than protecting individual tokens, except if those individuals are among the last of an endangered species. Indeed, if humans have eliminated top predators such as wolves and coyote, humans must take on the predator role in order to prevent prey—such as deer—from overshooting their resource base and causing much havoc and suffering in the process. Animal rights activists have challenged environmentalists for emphasizing species well-being, while ignoring the plight of individual organisms, which are sentient and possess a significant mode of interiority. Holmes Rolston III, who acknowledges the relative interior richness of wild animals, nevertheless sides with environmentalists who say that it makes no sense for humans to protect animals in the wild, especially when it comes to predation. Over thousands of years, the predator-prey relationship between, say, cougar and deer is responsible for making the deer more fleet of foot and the cougar more cunning.[17] Still, the experience of individual wild animals and the values of their communities are certainly worth taking into account when assessing how a proposed human intervention—whether constructing a new highway or clear cutting a forest—would impinge on wild animals. Given recent findings about the interiority of plants, environmentalists ought also to take the first-person perspective of individual plants and plant communities into account when studying a situation in which plants are harmed or threatened with destruction.[18] As an ecology for the future, integral ecology cannot expect that such considerations will be brought to the table any time soon, but holds open the possibility that further human ethical development will change this situation.

Just as it is presumptuous for anthropocentrists to treat nonhuman life as if it had value solely as raw material for human purposes, so it is misguided for anti-anthropocentrists to ignore that human beings represent a remarkable development in terrestrial evolution. Humans are special, in part because of their richly developed interiority made possible by their language-capacity, but interiority is not restricted to humans. Indeed, every holon has both an exterior and a corresponding interior.[19] Leopold inferred such interiority on the basis of what he saw in the eyes of the dying wolf. Humankind's linguistically enriched interiority makes possible not only the technological power needed to exploit nonhuman beings, but also the evaluative capacity to limit human actions out of respect for nonhuman Others. As Leopold remarks:

> For one species to mourn the death of another is a new thing under the sun. . . . But we, who have lost our [passenger] pigeons, mourn the loss. Had the funeral been ours, the pigeons would hardly have mourned us.[20]

The Slow Motion Inclusion of Values and Interiority

Recognition of the multifaceted character of environmental problems led the Ecological Society of America (ESA) to organize a plenary session at its August 2004 annual meeting to hear the chiefs of three major science-based federal agencies: the U.S. Department of Agriculture

Forest Service, the U.S. Geological Survey, and the National Oceanic and Atmospheric Administration. Indicating that he had been "humbled" by the daunting intricacy of environmental problems, Forest Service Chief Dale Bosworth in effect called for an integral ecology: "We need more than technical solutions to problems. We need to focus on the problem in its full dimension-its social and its regulatory and its political and its economic and its ecological dimensions."[21] Note that there are no references to the ethical, cultural, interpersonal, or aesthetic dimensions of such problems.

ESA's recent publication, *Ecological Science and Sustainability for a Crowded Planet* (ESSCP), offers important elements of an integral framework for ecology:

> Ecology is by its very nature an interdisciplinary science, making it impossible for any single ecologist to be well versed in the details of every relevant discipline, method, or instrument. Yet, it is increasingly obvious that ecologists must come together to help understand, solve, and anticipate the environmental issues facing our world. To do so, ecologists may need to think of themselves as entrepreneurs in a shifting and pressure-driven marketplace, where strategic collaborations and rapid responses are keys to scientific success. Our best chance to succeed in those efforts is to have a broadly inclusive approach to ecological research. This approach must include actively recruiting expertise beyond our discipline, as well as changing our culture to best foster the innovations we need.
>
> [. . . .] If successfully implemented, this new depth and breadth of ecological understanding, including its improved communication beyond the discipline, would allow ecologists to play an influential and eminently helpful role in decisions made at all levels that affect the sustainability of the biosphere.[22]

Although the document's title page encircles a drawing of the planet Earth with three phrases: "Anticipatory Research, Informed Decisions, Cultural Change," ESSCP accords a negligible role to culture and values. Discussion of culture is limited to changing the culture of natural science, currently characterized by method hegemony, in order to foster greater collaboration with other natural scientists, social scientists, businesspeople, and government officials. "Ethics" appears only once, in connection with moral rules that apply in using the data generated by others. The term "values" occurs in a paragraph encouraging scientists to provide "rigorous ecological knowledge" to religious groups that "have responded to emerging environmental concerns by linking values to an ethos of environmental stewardship."[23] To bring to fruition ESSCP's important vision of sustainability, greater attention must be given to studying personal and cultural factors—including values, worldviews, and religious beliefs—that play a role both in generating and resolving environmental issues, at all levels, from local to global.

A growing number of environmental scientists recognize that to be effective players in the hotly contested environmental arena, they must take into account perspectives other than those of natural science. Tainter, Allen, and Hoekstra call this "post-normal science."

> In post-normal science [. . .] data are insufficient, time is short, and because the stakes are high there is keen public interest and *conflicting values*. The findings of post-normal sciences are embedded in *a larger social framework*, in which the audiences consist of contending interest groups, and in which issues have more than one plausible solution. [My emphases.][24]

Post-normal environmental science, which emphasizes the importance of including stakeholders and alternative points of view, has much in common with integral ecology's attempt to incorporate findings from the domains of culture, society, and nature.

Another step toward integrating these three domains is taken by the editors of *Panarchy: Understanding Transformations in Human and Natural Systems*, the well-known ecologists, Lance H. Gunderson and C. S. Hollings. *Panarchy* calls on the rich conceptual model of resilience in complex adaptive systems to show how natural and social sciences can and must cooperate to address environmental problems.[25] In one essay, Frances Westley et al. argue that failure to understand the difference between ecological and social systems "helps to explain the fundamental lack of responsiveness or adaptability to environmental signals that characterize much of natural resource management."[26] Whereas space and time are key categories for understanding ecosystemic structures and patterns, "For social systems, we need to add a third dimension, which is symbolic construction of meaning."[27] This symbolic dimension makes possible capacities available only to humans: "the creation of a hierarchy of abstraction"; reflexivity in meaning; envisioning alternative futures; and externalizing "symbolic constructions in technology...."[28]

In another essay in *Panarchy*, "A Future of Surprises," Marco A. Jannsen not only discusses the interrelation of cultural domains, socioeconomic systems, and ecosystems, but also outlines a developmental model of culture (as worldviews) that has much in common with an integral ecology approach. According to Jannsen, the most prevalent U.S. worldviews are: *hierarchalism* (or conservatism), held by those who defer to authority in defining and solving environmental problems; *individualism*, affirmed by those who put faith in the power of unhindered markets to solve those problems; and *egalitarianism*, adhered to by those (including Greens) who claim that environmental problems can be solved primarily by reducing inequity across the board. Hierarchalism, individualism, and egalitarianism correspond in most important respects to traditional, modern, and postmodern developmental perspectives.[29]

Failure to differentiate among these three worldviews—traditional, modern, and postmodern—and to address adherents to each of them in rhetorically effective ways, is one reason that environmentalism is now widely viewed merely as an interest group, despite the fact that large percentages of Americans from all three worldviews uphold environmental values.[30] Indeed, these days we hear much about "the death of environmentalism."[31] In his *New York Times* column, "Geo-Greening by Example" (March 27, 2005), Thomas L. Friedman argues that environmentalism can be shaken from its current malaise by political leaders who present solutions to major eco-problems in terms consistent with the three major U.S. cultural worldviews: religious (conservative), neoconservatives, and Greens (corresponding to our categories of traditional, modern, and postmodern). Urging President George W. Bush to adopt major ecofriendly energy initiatives, Friedman says that doing so is

> . . . smart geopolitics. It's smart fiscal policy. It is smart climate policy. Most of all—it's smart politics. Even evangelicals are speaking out about our need to protect God's green earth. "The Republican Party is much greener than George Bush or Dick Cheney," remarked [Peter Schwartz, chairman of Global Business Network]. . . . Imagine if George Bush declared that he was getting rid of his limousine for an armor-plated Ford Escape hybrid, adopting a geo-green strategy and building *an alliance of neocons, evangelicals and greens to sustain it.* His popularity at home—and abroad—would soar. The

country is dying to be led on this. [My emphasis.]

Integral ecology maintains that ascribing a depth-dimension to the cosmos is justifiable in terms of current research into the sentience of animals and even plants, not to mention humans. Moreover, restoring such a depth dimension—in a way that can win the attention and consideration, if not yet the assent of moderns—is particularly important at this moment in human history, where traditional people are understandably suspicious of and resentful toward a modernity that holds their beliefs in contempt, but itself adheres to a cosmology that seems to invite despair.[32]

NOTES

1. On the issue of outstripping environmental resources, see William R. Catton, Jr., *Overshoot: The Ecological Basis of Revolutionary Change* (Urbana: University of Illinois Press, 1980). See the website *Die Off* for an extensive discussion and links to other discussions about short and long term problems associated with limits to the carrying capacity of the planet: http://dieoff.org/index.html. See also Jared Diamond, *Collapse: How Societies Choose to Fail or Succeed* (New York: Viking Press, 2004).
2. See Sean Esbjörn-Hargens and Michael E. Zimmerman, *Integral Ecology: Uniting Multiple Perspectives on the Natural World* (Boston: Integral Books, 2009).
3. See Kalevi Kull, "Biosemiotics in the Twentieth Century," *Semiotica*, 127, no. 1 (1999): 385–414. For an integrative view of semiotics, see Claus Emmeche, "The Biosemiotics of Emergent Properties in a Pluralist Ontology," in *Semiosis. Evolution. Energy: Towards a Reconceptualization of the Sign.*, ed. Edwina Taborsky, (Aachen, Germany: Shaker Verlag, 1999), 89–108.
4. Ken Wilber, *A Brief History of Everything*, 2nd ed. (Boston: Shambhala, 2001), 277.
5. *Ibid.*, 123.
6. *Ibid.*, quotations from 128 and 129 respectively.
7. *Ibid.*, 284.
8. *Ibid.*, all quotations from 263.
9. *Ibid.*, quotations from 288, 288, and 311 respectively.
10. Hans Jonas, *The Imperative of Responsibility: In Search of an Ethics for the Technological Age* (Chicago: University of Chicago Press, 1984).
11. Ken Wilber, foreword to I*ntegral Medicine: A Noetic Reader*, found at *Ken Wilber Online, http://wilber.shambhala.com/html/misc/integral-med-1.cfm.* Accessed February 16, 2006.
12. For an important account of how scientific truth claims are being challenged by alternative perspectives, see Joseph A. Tainter, T. F. H. Allen, and T. W. Hoekstra, "Energy Transformations and Post-Normal Science," *Energy* 31 (2006): 44–58.
13. Aldo Leopold, *A Sand County Almanac* (New York: Oxford University Press, 1949) is Leopold's best-known work. The ecophilosopher J. Baird Callicott has done much to promote Leopold's insights and value for contemporary environmental studies, especially Leopold's notion of the Land Ethic. See Callicott's *Beyond the Land Ethic: More Essays in Environmental Philosophy* (Albany: State University of New York Press, 1999) and his *In Defense of the Land Ethic: Essays in Environmental Philosophy* (Albany: State University of New York Press, 1989). He has also helped make many of Leopold's writings more available. See Aldo Leopold, *For the Health of the Land: Previously Unpublished Essays and Other Writings*, ed. J. Baird Callicott and Eric T. Freyfogle (Washington D.C.: Island Press, 1999); Aldo Leopold, *The River of the Mother of God And Other Essays*, ed. Susan L. Flader and J. Baird Callicott (Madison: University of Wisconsin Press, 1991); and J. Baird Callicott, ed. *Companion to a Sand County*

Almanac: Interpretive and Critical Essays (Madison: University of Wisconsin Press, 1987). Two worthwhile biographies of Leopold include Curt Meine, *Aldo Leopold: His Life and Work* (Madison: University of Wisconsin Press, 1991) and Marybeth Lorbiecki, *Aldo Leopold: A Fierce Green Fire* (Guilford, Conn.: Falcon, 2005).

14. Leopold, *Sand County Almanac*, 203.
15. *Ibid.*, 224–225.
16. *Ibid.*, 129–130.
17. See Holmes Rolston III, *Environmental Ethics* (Philadelphia: Temple University Press, 1998).
18. Alexandra Nagel, "Are Plants Conscious?" *Journal of Consciousness Studies*, 4, no. 3: 197, 215–230.
19. Panpsychism is the name given to the idea that the capacity for experience, however meager, is a basic feature of the universe. In recent times, Alfred North Whitehead was one of the most important exponents of this concept. Charles Birch and John B. Cobb, Jr., drew upon Whitehead's process philosophy in their book, *The Liberation of Life: From the Cell to the Community* (New York: Cambridge University Press, 1981). For a sophisticated defense of a version of panpsychism, see David Chalmers, *The Conscious Mind: In Search of a Theory* (New York: Oxford University Press, 1996).
20. Leopold, *Sand County Almanac*, 110. Thanks to Gus DiZerega for reminding us of this passage.
21. "Complexity 'Humbles' Environmental Chiefs," *The Oregonian*, Thursday, August 5, 2004, section C, 11.
22. Ecological Visions Committee Report to the Governing Board of the Ecological Society of America, *Ecological Science and Sustainability for a Crowded Planet* (Washington, D.C.: Ecological Society of America, April 2004) available online at: http://www.nau.edu/~envsci/sisklab/Lab%20Group%20Readings/EcologicalVisionsReport.pdf. Quotation is from 29.
23. *Ibid.*, 15.
24. Tainter, Allen, Hoekstra, "Energy Transformations," 45.
25. Lance H. Gunderson and C. S. Hollings, ed., *Panarchy: Understanding Transformations in Human and Natural Systems* (Washington, D.C.: Island Press, 2001).
26. Frances Westley, Steven R. Carpenter, William A. Brock, C. S. Holling, and Lance H. Gunderson, "Why Systems of People and Nature are not just Social and Ecological Systems," in *Panarchy*, 103-119. Quotation is from 119.
27. *Ibid.*, 119.
28. *Ibid.*, 105.
29. Marco A. Janssen, "A Future of Surprises," in *Panarchy*, 241–260. Other studies also come to the same conclusions, namely, that North American attitudes toward nature can be understood in part in terms of the threefold developmental levels, roughly premodern, modern, and postmodern or Green. See William M. Kempton, James S. Boster, and Jennifer A. Hartley, *Environmental Values in American Culture* (Cambridge: MIT Press, 1996), and Paul H. Ray

and Sherry Ruth Anderson, *The Cultural Creatives* (New York: Three Rivers Press, 2001).

30. See Kempton, Boster, and Hartley, *Environmental Values.*
31. See Ted Nordhaus and Michael Shellenberger, *Break Through: From the Death of Environmentalism to the Politics of Possibility* (New York: Houghton Mifflin, 2007). See also the symposium articles on Nordhaus and Shellenberger's original essay, "The Death of Environmentalism" in *Organization and Environment*, 19, no. 1 (March, 2006).
32. Recently, however, there has been a flood of books, authored by leading scientists, which attempt to reconcile science and religion. See for example Owen Gingrich, *God's Universe* (Cambridge: Belknap Press of Harvard University, 2006); Francis S. Collins, *The Language of God: A Scientist Presents Evidence for Belief* (New York: The Free Press, 2006); Stuart A. Kauffman, *Reinventing the Sacred: A New View of Science, Reason, and Religion* (New York: Basic Books, 2008).

From the Ground Up: Dark Green Religion and the Environmental Future[1]

Bron Taylor

Green and Dark Green Religion

Green religion as I use the term is a broad umbrella for every type of religious environmentalism, both those with deep roots in Western and Asian cultures, and more recent innovations that are emerging in the age of ecology. The types of religious environmentalism where practitioners and scholars affiliated with the world's most prevalent religious traditions seek to reveal and promote their putatively environmentally friendly dimensions, or develop such dimensions where they are believed to be missing or anemic, is not my present focus. This contemporary impulse to foster environmentally friendly religious ethics provides a backdrop for the exploration of the emergence, diffusion, characteristics, and types of a subset of green religion that I call *dark green religion*.

By dark green religion, I mean religion that considers nature to be sacred, imbued with intrinsic value, and worthy of reverent care. Dark green religion considers nonhuman species to have worth, regardless of their usefulness to human beings. Such religion expresses and promotes an ethics of kinship between human beings and other life forms. I use the title, "From the Ground Up," to focus on the intellectual roots of such spirituality by examining dark green religion within what I call *the environmentalist milieu*, namely, the contexts wherein environmentally concerned officials, movements, and individuals connect with and reciprocally influence one another.[2]

In recent decades, debates within the environmentalist milieu have raged over the relationships between religions, cultures, and the earth's living systems. Some assert that religious perceptions and beliefs have always been closely associated with natural phenomena, and that many religions originated in the worship of nature.[3] Others purport to find links between religious types and the specific natural habitats that they claim gave rise to them.[4] Still others take a decidedly reductionistic approach, asserting that religion is a byproduct of evolutionary processes.[5] Others have concluded, to the contrary, that religious beliefs and practices, including some forms of ritualizing, evolved in ecologically adaptive ways.[6] Such adaptive-functionalists provide a theoretical basis for the idea that religion can or does contribute to environmentally sustainable communities.

Better known are those who blame specific religions, or religion in general, for promoting worldviews that lead to environmental destruction.[7] Such criticisms and the responses they precipitate have led to a scholarly field most commonly called "religion and ecology." Using resources from existing religions, the field has been characterized by efforts to recover ideas that can be used to promote environmentally responsible attitudes and behavior. This work has been undertaken by religious thinkers, leaders, and practitioners, as well as by scholars who focus on specific traditions in an effort to help them become more environmentally friendly. The most impressive example of this scholarly enterprise was a series of conferences (and subsequently a book series) that unfolded between 1996 and 2004 on "Religions of the World and Ecology" organized by Mary Evelyn Tucker and John Grim, then professors in Bucknell's Religious Studies department. The Center for the Study of World Religions (CSWR) at Harvard University hosted the conferences with additional support from many other environmental, religious, and animal welfare groups, and the books were published by the CSWR and distributed by Harvard University Press.[8]

Put simply, the ferment has centered on how and under what circumstances religion can be "green." In other words, does it or can it assume forms that promote environmental sustainability?

Much evidence suggests a negative answer. Despite occasional and increasing expressions of environmental concern by practitioners of the world's major religious traditions, most of these traditions view their environmental responsibilities as, at most, one of a variety of ethical responsibilities. Clearly, *environmental* duties receive far less attention than what are considered to be *religious* duties and other, more pressing, ethical obligations. Nevertheless, diverse forms of green religion are emerging and going global in dramatic if nascent ways. Although both green religion and dark green religion have deep historical antecedents,[9] the growing strength and contemporary novelties make it possible to consider them both as new religious movements.

Nature as sacred is not new

Clarence Glacken's *Traces on the Rhodian Shore,* Donald Worster's *Nature's Economy,* and Lawrence Buell's *The Environmental Imagination,* all analyze important aspects of what can be called nature-as-sacred religions, namely, religions that consider nature itself to be inherently sacred, not only worthy of respect or reverence because it was created by a divine being. In different ways, these scholars illustrate that such religions have deep roots in longstanding organic and esoteric traditions in Western culture but note, however, that as scientific and ecological paradigms have shifted, so too have the forms of such spirituality.[10]

These and other scholars have analyzed the early and dramatic revival of such nature religions, often tracing this revival to eighteenth-century European romanticism, which influenced nature-related religious thinking in North America, which in turn reinforced and strengthened such movements in Europe.[11] Worster and Buell are among those scholars who have exposed the roots of what could be called the biocentric turn in ecological science and literature, namely, the turn toward values professing that nature has intrinsic or inherent value.

As Buell has shown, Henry David Thoreau is often regarded as a patron saint for such spirituality in America, casting a long shadow and influencing virtually all of the twentieth-century's most important environmentalist thinkers, including John Muir, John Burroughs, Aldo Leopold, Rachel Carson, Wendell Berry, Edward Abbey, Gary Snyder, and James Lovelock.[12] Indeed, both Thoreau and these progeny have assumed iconic status within the pantheon of saints favored among those who participate in contemporary nature religion.[13]

Four Types of Dark Green Religion

Here I will provide a few examples of four types of dark green spirituality that have been emerging since the first Earth Day. Just as "map is not territory," typological constructions are not meant to be exhaustively descriptive. Their boundaries are permeable and fluid. The key question is whether they have explanatory and heuristic power.[14]

The first two types of dark green religion I consider to be forms of Animism, one supernaturalistic and the other naturalistic. The third and fourth types I label Gaian Earth Religion, which similarly appears in two forms, one supernaturalistic the other naturalistic.[15] I use the expression Gaian Earth Religion as shorthand for holistic and organicist world views. The supernaturalistic form of Gaian Earth Religion I call "Gaian Spirituality," and the naturalistic form I label "Gaian Naturalism." All four of the above-mentioned types have fluid boundaries. They represent tendencies rather than uncomplicated, static, or rigid clusters of individuals and movements.

Animism is a term that most fundamentally reflects a perception that spiritual intelligences, or life-forces, animate natural entities and living things. Animistic perception is often accompanied by ethical beliefs about the kinds of relationships people have or should have with such beings or forces, or conversely, what behaviors should be avoided with regard to them. Animism may also involve communication or even communion with such intelligences or life forces. Such a worldview usually enjoins respect if not reverence for and veneration of such intelligences and forces.[16]

I parse my words carefully when speaking of spiritual intelligences or life forces. By using the term, *spiritual intelligences,* I seek to capture the beliefs of those for whom there is an immaterial, supernatural dimension to the Animistic perception. By the term, *life forces,* I refer to those who are agnostic or skeptical that any immaterial dimension underlies the life forces they perceive and with whom they seek understanding and connection. In both cases, Animism, as I configure the term, involves a shared perception that beings or entities in nature have their own integrity, ways of being, and even intelligence. With such Animism, we can, at least by conjecture and imagination, and sometimes through ritualized action and other practices, come to some understanding of these other life forces.[17]

Gaian Earth Religion, as I configure this construction, stands firmly in the organicist tradition, and takes the biosphere (or the universe) as a whole, and the complex internal relations of its constitutive parts and energetic systems, as the fundamental focus or object of understanding and respect.[18] Moreover, such a perspective takes the whole, usually as understood scientifically, but not always exclusively so, as a model. It thus defies the naturalistic fallacy argument in ethics, the assertion that one cannot logically derive value from fact, offering nature itself as sacred and thereby, at least implicitly, asserting that it contains both facts and values.

What I label *Gaian Spirituality* is avowedly supernaturalistic, perceiving the superorganism, whether the biosphere or the entire universe, to be an expression or part of God, or Brahman, or the Great Mystery, or by whatever name is used to symbolize the divine cosmos. This form is more likely to draw on deviant or nonmainstream or nonconsensus science for data to reinforce its generally pantheistic or panentheistic and holistic metaphysics. It is more open to interpretations commonly found in subcultures referred to as "New Age."

The form I call *Gaian Naturalism* represents a skeptical stance toward any supernaturalistic metaphysics. Its claims are more likely to be restricted to the scientific mainstream as a basis for understanding and promoting a holistic metaphysics. Yet, its proponents express awe and wonder when faced with the complexity and mysteries of life and the universe, relying on religious language and metaphors of the sacred, albeit not always self-consciously, when confessing feelings of belonging and connection to the energy and life systems in which they participate, live, and study.

Examples of Dark Green Religion

Exemplars of spiritual animism and Gaian spirituality include three thinkers whose spiritual paths involve serious encounters with Buddhism: Gary Snyder, Joanna Macy, and John Seed. All three also identify with deep ecology.

Gary Snyder is best known as a "beat" poet and one of the architects of bioregionalism, a social philosophy and branch of environmentalism that seeks to decentralize political decision-making processes so that they take place within the contours of differing ecological regions.[19] In an interview with me in 1993, Snyder called himself a "Buddhist-Animist," meaning that he thinks that the world is full of spiritual intelligences. Most children and many indigenous peoples who live close to nature have similar perceptions, Snyder asserted, but for those who live in degraded habitats divorced from nature, such perceptivity is easily lost and must be self-consciously rekindled.

Snyder's solution is to encourage a bioregional "reinhabitation" of particular places: by going "back to the land," people can recover their ability to hear nature's multivocal, sacred voices. Snyder and others in his intentional community, located in the foothills of the Sierra Nevada Mountains for a generation, have drawn upon many traditions in their experimentations with nature-related ritual.[20]

Joanna Macy and John Seed have followed a kindred religious path, but unlike Snyder, they have labored to spread globally the ritual processes they developed to reconnect people to the earth and its inhabitants. Their best known ritual is the Council of All Beings, which has inspired further experimentation with nature-focused ritualizing. The "sacred intention" of these rituals is to reawaken lost understandings of spiritual realities, which they believe animate nature in its many expressions.[21]

The experience participants have during the Council varies. Some report being possessed by and speaking for the spirits of nonhuman entities. This kind of experience seems to fit into what I call spiritual animism. During the Council other participants speak for DNA or energy pulses permeating the universe or of the pain felt by Gaia from mining or the polluting of her waters, an experience that seems to fit what I call Gaian spirituality. In both cases the participants have what most would consider to be a *religious* experience.

Speaking for a nonhuman life form is for other Council participants more an act of moral imagination than an experience of being called by a spiritual intelligence or a feeling of connection to a divine universe. For them, the Council is ritualized performance art in which participants act out what they surmise it must feel like for the earth, or some earthly entity, which is being badly mistreated by human beings. This understanding, depending on the form of expression, might be aptly labeled naturalistic animism or Gaian naturalism.

While naturalistic animism involves disbelief that some parallel spiritual world animates nonhuman natural entities, it nevertheless affirms an experience of kinship with and ethical concern for nonhuman life, and sometimes a felt communion with it.

According to Donald Worster, this felt kinship and the biocentric ethics that often accompanies it can be grounded in evolutionary theory and was expressed by Charles Darwin himself:

> If we choose to let conjecture run wild, then animals, our fellow brethren in pain, diseases, death, suffering and famine—our slaves in the most laborious works, our companions in our amusements—they may partake [of] our origin in one common ancestor—we may be all netted together.[22]

Darwin believed that this kinship ethic can be deduced through reflection on an awareness of a common ancestor and kinship with other animals who suffer and face challenges, as do we. This is a form of empathetic moral imagination from which understanding and communion arise. Animism understood in this way can be entirely independent of metaphysical speculation or supernaturalistic assumptions.

Through interviews with both radical and pragmatic environmentalists, and in a wide range of environmental literatures, examples of such feelings and perceptions can be found, usually among those who endorse evolutionary theory's supposition of a common ancestor. Some ethologists, for example, articulate such a view. Naturalistic animism, indeed, is not uncommon among those who study primates, elephants, and other animals, especially mammals. For example, Katy Payne, an acoustic biologist, has scrutinized elephant communication, concluding that human-elephant communication is possible for attentive humans.[23] Increasing numbers of scientists are also finding communicative and affective similarities among humans and other animals.[24]

The biologist/ethologist Marc Bekoff is a well-known proponent of such naturalistic animism, speaking and publishing widely, including recently in *Minding Animals: Awareness, Emotions, and Heart.*[25] The famous primatologist, Jane Goodall, wrote the foreword to this book and subsequently coauthored with him *The Ten Trusts: What We Must Do to Care for the Animals We Love.* In a section entitled "the power of eyes," Goodall recalled a story about a chimpanzee named JoJo who was orphaned young and had grown unfamiliar with chimpanzee ways after living alone and growing into adulthood in a cage. When eventually taken to a zoo enclosure, JoJo was threatened by more aggressive chimpanzees from whom he fled in terror, falling into a water-filled moat, where he began to drown. A visitor, at the risk of his own life, ignored the threatening chimpanzees, jumped into the enclosure, and pulled JoJo out of the water. According to Goodall, when asked what made him do it, the visitor answered, "I happened to look into his eyes, and it was like looking into the eyes of a man. And the message was, 'Won't *anybody* help me?'" Goodall commented,

> I have seen that appeal for help in the eyes of so many suffering creatures . . . All around us, all around the world, suffering individuals look toward us with a plea in their eyes, asking us for help.
>
> And if we dare to look into those eyes, then we shall feel their suffering in our hearts. More and more people have seen that appeal and felt it in their hearts. All around the world there is an awakening of understanding and compassion, an understanding that reaches out to help the suffering animals in their vanishing homelands. . . . Together we can bring change to the world, gradually replacing fear and hatred with compassion and love. Love for all living beings. [26]

Earlier in *The Ten Trusts*, Goodall and Bekoff wrote that our obligation is to open our minds in humility to animals and learn from them. Through their books and lectures Goodall and Bekoff are indeed promoting an awareness and openness to an empathetic interspecies understanding.

Goodall has become the world's foremost missionary promoting naturalistic animism, although she also believes in reincarnation and God, at least as understood in some pantheistic sense.[27] In worldwide lecture tours, she promotes her animistic nature spirituality empowered by her designation as a United Nations Ambassador for Peace in 2002. She is not, however, the only one whose nature-related spirituality has been shaped by felt understandings and communication with nonhuman beings, developed through ordinary observational capacities rather than gained by mystical religious epiphany.[28]

L. Freeman House provides another example of both spiritual and naturalistic animism. As a friend of Snyder's and a fellow pioneer of bioregional spirituality and politics, House was part of the countercultural back-to-the-land movement in Northern California. While living there in a remote coastal valley in Northern California, House became involved with and felt closely connected to salmon, whose populations had declined dramatically in many Pacific watersheds due to dam building, logging, and erosion. He and the other settlers who had arrived in this watershed during the 1960s labored to protect and bring the salmon back to viable populations. In a remarkable 1974 essay entitled "Totem Salmon," he described the cultural, spiritual, and material significance of salmon to the indigenous people of the North Pacific Rim where aboriginal peoples had ceremonies to ensure that the salmon would take no offence when caught. Such practices, he averred, were based on:

> . . . [the] notion that conscious spirit resides in all plants and animals. [Therefore,] the Salmon is always perceived as a person living a life similar to that of the people who catch it. Therefore, before it is safe to eat any plant or animal it is necessary to assure the creature that *there is no desire to offend*. Thus the ceremonies . . . have the practical effect of assuring the continuity of both species, salmon and human.[29]

House shared the animistic spirituality that he believed characterizes the indigenous peoples he had studied.[30] He also signaled the possibility of a *naturalistic* animism when he argued that the salmon speak to humans *practically* about appropriate lifeways. To paraphrase: salmon speak to humans, if only by their disappearance.[31] Such communication may be considered naturalistic animism.

It is useful to look at others who might be considered in a similar light. The environmental philosopher Paul Taylor, for example, argued in an important 1986 book that all beings who are "subjects of a life" have interests that ought to be respected,[32] and the animal rights philosopher, Tom Regan, whose writings inspired such activists, grounded his theories in his own affective and personal connection with sentient animals.[33] Regan has even invented a spiritual practice by urging animal rights activists to select a "totem animal," based on ones they could have helped at some point but failed, suggesting that they draw on that animal's strength whenever their passion wanes for the animal rights cause. Perhaps this also could be considered to be a form of naturalistic animism, for it involves a belief that one can empathetically understand the feelings and protect the interests of nonhuman animals.[34]

The American ecologist Aldo Leopold provides an example of a bridge between naturalistic animism and Gaian spirituality. In 1949 Leopold's posthumously published essay entitled "Thinking Like a Mountain" was published in *A Sand County Almanac,* subsequently becoming a well-known sacred story to many environmentalists.[35] The essay described an epiphany Leopold had after he and his comrades shot a she-wolf and he looked into her eyes while she was dying. Although he had once helped exterminate the species with both pen and gun, through that eye-to-eye contact he realized that the wolf had value for her own sake and value also to the mountain (a metaphor for nature herself), that superceded human interests. There are many similar examples wherein a callous killing has led to connection, understanding, and communion between a human and nonhuman being, leading to a life dedicated to animal or environmental activism.

Leopold's awakening involved more than simply appreciating the value of an individual animal or its species, however, it contributed decisively to his ethical holism. Leopold stood firmly in the organic tradition in a way that regarded the natural world as sacred. That regard was enhanced by the ecological science prevalent during his time:

> The land is one organism . . . . The outstanding discovery of the twentieth century is . . . the complexity of the land organism. If the land mechanism as a whole is good, then every part is good, whether we understand it or not.[36]

> Possibly, in our intuitive perceptions, which may be truer than our science and less impeded by words than our philosophies, we realize the indivisibility of the earth—its soil, mountains, rivers, forests, climate, plants, and animals, and respect it collectively not only as a useful servant but as a living being, vastly less alive than ourselves in degree, but vastly greater than ourselves in time and space. . . .[37]

Curt Meine's biography of Leopold recorded Leopold's deep spiritual connection to the earth's living systems, along with his profound sense of their sacredness, noting that late in Leopold's life his youngest daughter, Estella, asked him directly about his belief in God. She later recalled,

> He replied that he believed there was a mystical supreme power that guided the universe but to him this power was not a personalized God. It was more akin to the laws of nature. He thought organized religion was all right for many people, but he did not partake of it himself, having left that behind him a long time ago. His religion came from nature, he said.[38]

With regard to the value of and possibility of communication with animals, there was naturalistic animism in Leopold's thinking. With regard to his holistic view of ecological systems and of the universe as a whole, Leopold's perceptions seem to reflect Gaian spirituality.

It was James Lovelock, of course, who resurrected Gaia and inserted the ancient Greek god of the earth into contemporary environmental discourse. Articulating the now famous "Gaia theory," Lovelock argued that the biosphere should be understood as a self-regulating organism that maintains the conditions necessary for the various individual species and organisms that constitute it.[39] When published in 1979 he understood the theory in purely scientific terms, and as such, it well represented what I call Gaian naturalism. Lovelock was surprised that the majority of the mail he received expressed interest in the theory's spiritual or religious dimensions. A portion of that response clearly sought to understand Gaia, the earth system organism, as a spiritual system or being, taking the theory more as a Gaian spirituality trope than Lovelock, who has staunchly maintained his agnosticism regarding metaphysical matters. Yet, he clearly enjoyed and appreciated, even if with some bemusement, those who refer to the Gaian system in explicitly religious terms. A good example of his sentiments can be found when, in 2001, he reflected on a speech given by President Václav Havel of the Czech Republic.

> When he was awarded the Freedom Medal of the United States . . . [Havel] reminded us that science had replaced religion as the authoritative source of knowledge about life and the cosmos but that modern reductionist science offers no moral guidance. He went on to say that recent holistic science did offer something to fill this moral void. He offered Gaia as something to which we could be accountable. If we could revere our planet with the same respect and love that we gave in the past to God, it would benefit us as well as the Earth. Perhaps those who have faith might see this as God's will also.[40]

Here, in a subtle way, Lovelock expressed his appreciation for and an affinity with Havel, who found a reverence for the earth through the holistic science represented by the Gaia theory, apparently having left behind traditional theism. Lovelock also signaled, to use the terms of discourse in the current analysis, that there is room for agreement among Gaian naturalists, traditional theists who revere an earth they believe God created, and devotees of Gaian spirituality, namely, those who consider the earth organism itself as sacred or divine. And in an essay reflecting on his own "Gaian Pilgrimage" and entitled after it, Lovelock expressed his deep feelings of belonging to and reverence for the earth's living systems.[41]

Lovelock's fusion of Gaia as both worthy of reverence and reverent care, with an acknowledgement of her mortality, may require some effort to fit into definitions of religion that require a belief in immortal divine beings. But it is a wonderful example of the newer forms of religious production I call Gaian naturalism.[42] The felt sense of connection and belonging expressed by Lovelock is commonly expressed by many environmentalists, whether they are traditionally religious or self-consciously atheistic or agnostic.[43] Combined with the environmental concern that is found in the above quotation, as well as in another work, *Healing Gaia: Practical Medicine for the Planet*, we see why Lovelock's thinking could be considered an example of dark green religion.[44]

There are other thinkers and social movements like Lovelock's, that could be included as good examples of Gaian naturalism, among them the World Pantheist Movement (WPM). Originally named the Society for Scientific Pantheism, in 2006, the group's website began with an epigraph attributed to Albert Einstein:

> A knowledge of the existence of something we cannot penetrate, of the manifestations of the profoundest reason and the most radiant beauty—it is this knowledge and this emotion that constitute the truly religious attitude; in this sense, and this alone, I am a deeply religious man.[45]

The website continued:

> Is Nature your spiritual home? Do you feel a deep sense of peace and belonging and wonder in the midst of nature, in a forest, by the ocean, or on a mountain top? Are you speechless with awe when you look up at the sky on a clear moonless night and see the Milky Way strewn with stars as thick as sand on a beach?

The next section asked, "Why do we need a spirituality of nature?" and answered, "Most people have a sense that there is something greater than the self or than the human race. The WPM's naturalistic reverence for nature can satisfy this need, without sacrificing logic or respect for evidence and science."

The site also encouraged social and environmental action, urging visitors to endorse the Earth Charter and fight global warming and economic inequality.[46] Moreover, it listed as "honorary members," a number of individuals who could also be considered exemplars of Gaian Naturalism, including Lovelock; Dr. David Suzuki, the noted Canadian science newscaster; Carl Sagan, the late astrophysicist and television celebrity who promoted widely a sense of wonder for the universe; and Ursula Goodenough, a professor of biology at Washington University in St. Louis, and an effective proponent of scientific, "religious naturalism."

Conclusion ~ A Dark Green Religious Future?

For more than a generation some scholars closely affiliated with the world's dominant religious traditions, at least those considered as "world religions," have labored to turn them in more environmentally friendly directions. This has often involved an expanded understanding of nature as sacred, or at least, a belief that protecting nature is a religious duty.

My analytic focus here, however, has been on "dark green religion," a form of nature-related spirituality that shares the impulse toward environmental concern but that also considers nature and its denizens sacred in and of themselves. With such religion, ethical obligations to nature are direct rather than only arising indirectly as a means to promote human well-being. Such nature spirituality is decreasingly tethered and sometimes entirely independent of the world's major religious traditions.[47]

Many additional examples of each type of dark green religion could be provided, as well as of those individuals and movements that cross the fluid lines between these types. Such religion is beginning to exercise influence in critically important sectors of the global environmental intelligentsia. It may even contribute eventually to the emergence of a new, civic earth religion.[48]

If the trope of "dark green religion," and the fourfold typology outlined here have heuristic value, then readers acquainted with environmental literature, movements, and politics will be able to fashion their own apt examples. As these new forms of nature religion spread globally and increase in influence, the examples will multiply, and their significance in global environmental politics will intensify.

NOTES

1. An in-depth analysis based on the framework introduced in this article will be published as *Dark Green Religion: Nature Religion and the Planetary Future* by the University of California Press in 2009.
2. This phrase was inspired by and adapted from Colin Campbell's notion of the "cultic milieu," by which he meant the Western countercultures in which socially deviant, countercultural knowledges, both spiritual and scientific/quasi-scientific, are brought together by their carriers and proponents, to incubate and cross-fertilize. The milieu is remarkably receptive to the ideas of the others in resistance to the cultural mainstream. His 1972 article, "The Cult, the Cultic Milieu and Secularization," is reprinted in *The Cultic Milieu: Oppositional Subcultures in an Age of Globalization*, eds. Jeffrey Kaplan and Heléne Lööw (Walnut Creek, Calif.: Altamira/Rowman and Littlefield, 2002), 12–25.
3. For especially influential examples, see the work of Max Müller in Jon Stone, ed., *The Essential Max Müller* (New York: Palgrave/Macmillan, 2002) and Sir James George Frazer, *The Golden Bough: A History of Myth and Religion* (London: Chancellor Press, 1994).
4. John Snarey, "The Natural Environment's Impact on Religious Ethics: A Cross-Cultural Study," *Journal for the Scientific Study of Religion* 35, no. 2 (1996): 85–96.
5. For a good introduction to such discussions see Herbert Burhenn, "Ecological Approaches to the Study of Religion," *Method and Theory in the Study of Religion*, 9, no. 2 (1997): 111–26; and Gustavo Benavides, "Ecology and Religion," in the *Encyclopedia of Religion and Nature*, ed. Bron Taylor (London & New York: Continuum International, 2005), 1:548–54.
6. For example, see Roy Rappaport, *Pigs for the Ancestors: Ritual in the Ecology of a New Guinea People* (New Haven, Conn.: Yale University Press, 1968) and J. Stephen Lansing, *Priests and Programmers: Technologies of Power in the Engineered Landscape of Bali* (Princeton, N.J.: Princeton University Press, 1991).
7. Most famously, Lynn White, Jr. but as influential within the environmental milieu has been Paul Shepard, who traces environmental decline to agricultures including salvation religions, an argument well summarized in *Coming Home to the Pleistocene* (San Francisco: Island Press, 1998); for Asian and indigenous religions who, as some have suggested, provide more fertile ground for environmental ethics than Abrahamic religions, see J. Baird Callicott and Roger T. Ames, ed., *Nature in Asian Traditions of Thought: Essays in Environmental Philosophy* (Albany: State University of New York Press, 1989). For strong challenges to assertions of the environmental beneficence of Asian and indigenous traditions, see Yi Fu Tuan, "Discrepancies Between Environmental Attitude and Behaviour: Examples From Europe and China," *The Canadian Geographer* 12 (1968): 176–91; Ole Bruun and Arne Kalland, *Asian Perceptions of Nature: A Critical Approach* (London: Curzon Press, 1995); Ian Harris, "Buddhist Environmental Ethics," *Religion* 25 (1995): 199–211; and Shepard Kretch, *The Ecological Indian: Myth and History* (New York: Norton, 1999). For an em-

pirical study to test Lynn White's thesis, see James Proctor and Evan Berry, "Social Science on Religion and Nature" in the *Encyclopedia of Religion and Nature*, 2:1571–76.

8. For a broader overview of the emergence of the religion and ecology field, see Bron Taylor, "Religious Studies and Environmental Concern," in *Encyclopedia of Religion and Nature*, 2:1373–79. Also available at http://www.religionandnature.com/ern/sample/Taylor--ReligiousStudiesEnviConcern.pdf.
9. Roderick Frazier Nash, *Wilderness and the American Mind*, 4th ed. (New Haven: Yale University Press, 2001); Catherine L. Albanese, *Nature Religion in America: From the Algonkian Indians to the New Age* (Chicago: Chicago University Press, 1990); Stephen Fox, *The American Conservation Movement: John Muir and His Legacy* (Madison: University of Wisconsin Press, 1981).
10. Clarence Glacken, *Traces on the Rhodian Shore: Nature and Culture in Western Thought From Ancient Times to the End of the Eighteenth Century* (Berkeley: University of California Press, 1967); Donald Worster, *Nature's Economy: A History of Ecological Ideas*, 2d ed. (Cambridge: Cambridge University Press, 1994); Lawrence Buell, *The Environmental Imagination: Thoreau, Nature Writing, and the Formation of American Culture* (Cambridge: Belknap/Harvard University Press, 1996).
11. For a starting point, see Joan Steigerwald, "Romanticism in European History" (2:1419–22) and Tilar J. Mazzeo, "Romanticism–American" (2:1424–26), in the *Encyclopedia of Religion and Nature*. For an analysis focused on North America, see Bron Taylor, "Religion and Environmentalism in North America and Beyond," *Oxford Handbook on Religion and Ecology*, ed. Roger S. Gottlieb (Oxford.: Oxford University Press, 2006).
12. Buell speaks of the "personification of nature" used by some environmentalist writers, but does not seem to connect this to animistic spiritual perception, nor does he mention theorists such as Stewart Guthrie who in, *Faces in the Clouds: a New Theory of Religion* (Oxford & New York: Oxford University Press, 1993), view animistic personification as the root of religion.
13. "The terms, *nature religion* or the plural, *nature religions*, are most commonly used as proxies for religious perceptions and practices that, despite substantial diversity, are characterized by a reverence for nature and consider nature sacred . . . . The term, *nature religion*, which began to be employed regularly within religious subcultures the time of the first Earth Day celebration in 1970, increasingly is used to represent and debate such 'nature-as-sacred' religion in both popular and scholarly venues." From Bron Taylor, "Ecology and Nature Religions," in *Encyclopedia of Religion*, ed. Lindsay Jones (New York: MacMillan, 2005), 4:2661–66.
14. The expression "map is not the territory" was coined by Eric Bell, popularized by Alfred Korzybski, and borrowed by J. Z. Smith as the title of his important book, *Map Is Not Territory: Studies in the History of Religions* (Chicago: University of Chicago Press, 1978).
15. Some scholars insist that belief in a nonmaterial divine being or beings is an

essential characteristic of religion, a view that I challenge in "Exploring Religion, Nature, and Culture." *Journal for the Study of Religion, Nature, and Culture 1*, no. 1 2007): 5-24, also at http://www.religionandnature.com/journal/sample/Taylor--JSRNC(1-1).pdf.

16. The suggestion that animism involves the worship of natural entities is often a projection based upon Western religious assumptions that have more to do with how humans relate to high god(s) than how they relate to spiritual intelligences in nature. Veneration or "profound respect or reverence" (www.dictionary.com) is a word that involves less Western projection. When it comes to animism, veneration is a more common posture than worship, as I understand the phenomena that the term seeks to capture.
17. Some scholars eschew the word animism, given its genesis as a term invented to distinguish, they say, superior monotheistic religions from inferior, primitive, African ones. For a useful introduction to the term animism, see David Chidester, "Animism" (1:78–81) and Graham Harvey, "Animism: A Contemporary Perspective" (1:81–83) in the *Encyclopedia of Religion and Nature*. See also Graham Harvey, *Animism: Respecting the Living World* (New York: Columbia University Press, 2006).
18. By "organicism" I mean not only the belief that the biosphere and universe are analogous to a biological organism, but also, that this organism is somehow sacred and worthy of reverence. Taken together, the books by Glacken and Worster (see note 10 above) provide a comprehensive survey of organicism in Western history.
19. For perhaps the earliest countercultural statement, see Gary Snyder, *Turtle Island* (New York: New Directions, 1969). For an early manifesto, see Raymond Dasmann and Peter Berg, "Reinhabiting California," in *Reinhabiting a Separate Country*, ed. Peter Berg (San Francisco: Planet Drum, 1978), 217–20. For an analysis of the movement, see Bron Taylor, "Bioregionalism: An Ethics of Loyalty to Place," *Landscape Journal* 19, nos. 1–2 (2000): 50–72.
20. These quotes come from my interview with Gary Snyder, 1 June 1993, Davis, California.
21. See the following entries: Joanna Macy, "Council of All Beings" (1:425–29), Craig S. Strobel, "Macy, Joanna" (2:1019–20); John Seed, "Re-Earthing" (2:1354–58); and Bron Taylor, "Snyder, Gary—and the Invention of Bioregional Spirituality and Politics" (2:1562–567) in the *Encyclopedia of Religion and Nature*. See also the discussion of the Council in B. Taylor, "Earth First!'s Religious Radicalism," in *Ecological Prospects: Scientific, Religious, and Aesthetic Perspectives*, ed. Christopher Key Chapple (Albany: State University of New York Press, 1994), 185–209.
22. From Charles Darwin's 'Notebooks on Transmutation,' quoted by Donald Worster in *Nature's Economy*, 180 (see footnote 10).
23. See Katy Payne, *Silent Thunder: In the Presence of Elephants* (New York: Penguin, 1999).
24. See Marc Bekoff, Colin Allen, and Gordon Burghardt, ed. *The Cognitive Ani-*

mal: Empirical and Theoretical Perspectives on Animal Cognition (Cambridge: MIT Press, 2002).

25. Mark Bekoff and Jane Goodall, *Minding Animals: Awareness, Emotions, and Heart* (Oxford & New York: Oxford University Press, 2002).
26. See Jane Goodall and Marc Bekoff, *The Ten Trusts: What We Must Do to Care for the Animals We Love* (San Francisco: Harper San Francisco, 2003), 169–71. See also Goodall's *Reason for Hope* (New York: Warner, 1999), 250, for another story about JoJo.
27. For passages with variously animistic and theistic dimensions, in which she expresses eclectic religious beliefs in a way common today, see Goodall, *Reason for Hope*, 11, 39, 72–73, 172–73, 199–200, 251, 266–69. In these passages, she clearly expresses what I have called "spiritualities of connection" to the earth, as well as a mystical Mother Earth spirituality (at 251) and belief in reincarnation (at 264).
28. Many more examples of people being moved by and connected to nonhuman beings by a perception of communication through eye contact could be noted. For another example of human/animal contact, see Paul Watson, *Seal Wars: Twenty-Five Years on the Front Lines with the Harp Seals* (Buffalo, N.Y.: Firefly Books, 2002), 78. For an example from a famous nature photographer, see the introduction in Frans Lanting, *Eye to Eye: Intimate Encounters with the Animal World*, ed. Christine Eckstrom (Köln, Germany: Taschen, 1997), 14–15.
29. See Freeman House, "Totem Salmon," in Van Andruss, Christopher Plant, Judith Plant, and Eleanor Wright, ed., *Home!: A Bioregional Reader* (Philadelphia: New Society, 1990), 65–72.
30. This was clear during a 3 June 1993 interview I had with House in Petrolia, California, in which he noted approvingly, "Further, the spirits of plants and animals were considered immortal."
31. House concluded that salmon are also telling us, among other things, "let's get serious about this business of coevolution." See House, "To Learn the Things We Need to Know: Engaging the Particulars of the Planet's Recovery," in *Home!: A Bioregional Reader*, 111–20. Later House published a book about his community's efforts to save the salmon entitled *Totem Salmon: Life Lessons from Another Species* (Boston: Beacon Press, 1999).
32. Paul W. Taylor, *Respect for Nature: A Theory of Environmental Ethics* (Princeton, N.J.: Princeton University Press, 1986).
33. Tom Regan and Peter Singer are the two chief proponents, respectively, of an animal rights and an animal liberationist perspective. See Tom Regan and Peter Singer, ed., *Animal Rights and Human Obligations*, 2d ed. (Englewood Cliffs, N.J.: Prentice Hall, 1989). For Regan's classic see *The Case for Animal Rights* (Berkeley: University of California Press, 1983).
34. Tom Regan, interview with the author, 14 February 2003, Fresno, California. For another type of naturalistic animism, see Cleve Backster, *Primary Perception: Biocommunication with Plants, Living Foods and Human Cells* (Anza, Calif.: White Rose Millennium Press, 2003).

35. Rick McIntyre, ed., *War against the Wolf: America's Campaign to Exterminate the Wolf* (Osceola, Wis.: Voyageur Press, 1995), 187–91, cf. 321–27. (Editor's note: see also Zimmerman's essay in this volume.)
36. For the original, see Aldo Leopold, *A Sand County Almanac with Essays from Round River* (Oxford: Oxford University Press, 1949); this quote is from "Essays from Round River," which appears only in the enlarged edition but not the original one (New York: Ballantine Books, 1966), 190.
37. See Susan Flader and J. Baird Callicott, ed., *The River of the Mother of God and Other Essays by Aldo Leopold* (Madison: University of Wisconsin Press, 1991), 95.
38. Curt Meine, *Aldo Leopold: His Life and Work* (Madison: University of Wisconsin Press, 1988), 506.
39. James Lovelock, *Gaia: A New Look At Life on Earth,* rev. ed. (Oxford & New York: Oxford University Press, 1995). Lovelock published his most complete articulation of the theory in *The Ages of Gaia: A Biography of Our Living Earth* (New York: Norton, 1988).
40. James Lovelock, "Gaian Pilgrimage," in the *Encyclopedia of Religion and Nature,* 1: 685.
41. Ibid., 683–85.
42. Others include J. Baird Callicott "Natural History as Natural Religion," in the *Encyclopedia of Religion and Nature,* 2:1164–69. Of course, such spirituality is not entirely new, as Ursula Goodenough shows in "Religious Naturalism," in the *Encyclopedia of Religion and Nature,* 2:1371–73. See also Goodenough, *The Sacred Depths of Nature* (New York & Oxford: Oxford University Press, 1998). For examples that sometimes flirt with theism, see Loyal Rue, *Everybody's Story: Wising Up to the Epic of Evolution* (Albany: State University of New York Press, 2000); and Brian Swimme and Thomas Berry, *The Universe Story: From the Primordial Flaring Forth to the Ecozoic Era: A Celebration of the Unfolding of the Cosmos* (San Francisco: Harper Collins, 1992).
43. For other references on nature religions and the difficulties that inhere to deciding how to define them, see Bron Taylor, "Ecology and Nature Religions," in Lindsay Jones, ed., *Encyclopedia of Religion* (New York: MacMillan, 2005), 4:2661–66; "Earth and Nature-Based Spirituality (Part I): From Deep Ecology to Radical Environmentalism," *Religion 31,* no. 2 (2001): 175–93; and "Earth and Nature-Based Spirituality (Part II): From Deep Ecology to Scientific Paganism," *Religion 31,* no. 3 (2001): 225–45.
44. James Lovelock, *Healing Gaia: Practical Medicine for the Planet* (New York: Harmony, 1991). His most recent book provides further evidence for my argument: see Lovelock's *The Revenge of Gaia: Earth's Climate Crisis and the Fate of Humanity* (New York: Basic Books, 2006).
45. All passages attributed to the World Pantheist Movement website were accessed in February 2006 at www.pantheism.net; see also Paul Harrison, "World Pantheist Movement" in the *Encyclopedia of Religion and Nature,* 2:1769–70. For additional examples, see David Suzuki and Peter Knudtson, *Wisdom of the*

Elders: Honoring Sacred Native Visions of Nature (New York: Bantam, 1992), especially 227, where the authors quote a statement issued in the early 1990s by a group of prominent scientists (including Stephen Jay Gould, Hans Bethe, Stephen Schneider, Carl Sagan, and Peter Raven) proclaiming, "As scientists many of us have had profound personal experiences of awe and reverence before the universe. We understand that what is regarded as sacred is more likely to be treated with care and respect. Our planetary home should be so regarded. Efforts to safeguard and cherish the environment should be infused with a vision of the sacred."

46. For the Earth Charter's religiously plural construction of nature as worthy of reverent care, see Steven C. Rockefeller, "Earth Charter," in the *Encyclopedia of Religion and Nature*, 1:516–18, and for the document and supportive website, see www.earthcharter.org.
47. If animistic perceptions and religiosities are themselves world religions, as proponents such as Gary Snyder suggest, then what we mean by "world religions" needs revision. I am not convinced that this is necessary, for while animistic perceptions and spirituality are longstanding and widespread, even in the modern world and in new ways, if I am correct in my own observations and analysis, they are nevertheless largely local phenomena. There is, therefore, no analytic advantage in insisting that they be considered "world religions."
48. See Bron Taylor, "A Green Future for Religion?" *Futures Journal* 36, no. 9 (2004): 991–1008.

Cultural Readings of the "Natural" World

Michael Jackson

In October 1993 I began ethnographic fieldwork on southeast Cape York, Australia, with the object of exploring the genealogies and entailments of competing views of the environment held by property developers, eco-activists, and local Aboriginal people. On a brief visit to the region in 1988, I had witnessed confrontations between Greens and road-makers along the newly bulldozed four-wheel drive track north of Daintree. I was also well aware of the dismay among conservationists when Kuku-Yalanji argued for, rather than against the road, claiming a need for better communications between their isolated settlements, even though this might lead to further European incursions into areas used for camping, hunting and gathering, as well as traditional burial places and sacred sites. The Greens' consternation reflected their commitment to a pervasive "myth of primitive ecological wisdom"[1] that assumes that Aboriginal people live in harmony with, and are closer to, nature than modern Europeans. Arguing against this essentially racist notion, as well as its corollary—that if Aboriginal people seem to abet environmental vandalism it is because they have lost their traditional culture—anthropologist Chris Anderson has pointed out that it was local politics, not culture or nature, that led the most vocal and powerful Kuku-Yalanji group to welcome the road because it stood to gain material benefits and consolidate its power in the mission settlement of Wujal Wujal through better access to the outside world.[2] In 1993, my wife and I would discover that the same local politics governed

Kuku-Yalanji discourse on a proposed Native Title claim to the Daintree rainforests.

However, my emphasis in this essay is not Kuku-Yalanji internal politics per se, but the life-world of the politically marginal Kuku-Yalanji family with whom my wife and I, and our two-year old son, lived for a year in 1993–1994.

As a result of the Queensland government's assimilationist policies, our host family, the Olbars, were forcibly moved in 1970 from their traditional land to a Lutheran mission settlement at Wujal Wujal on the Bloomfield River. However, in 1992 they left the settlement to take up residence on a parcel of their former land that had been purchased for them by the Aboriginal and Torres Strait Island Commission (ATSIC).

Every afternoon we would accompany our hosts on expeditions to the nearby beach at Weary Bay or to the Bloomfield rivermouth to forage or fish. Unlike my wife, Francine, I lacked the patience for fishing, and often preferred to stroll along the beach, listening to the wind in the casuarinas, observing the stingrays moving like cloud shadows beneath the waves, or watching container ships inching as slowly as clock hands along the horizon and Torres Strait pigeons flying in from the open sea to feed in the forest. Entranced by what I experienced as the pristine nature of the environment, I commented to Mabel, our host, that it was very beautiful.

My remark was immediately rebutted by a very pragmatic set of observations.

"This is my *bubu*, my country," Mabel said. And she described the green turtles offshore, the bush yams and *dakay* (mud clams) in the scrub, and said that she hoped it would not be long before she and her family would reclaim all their land, and live undisturbed by outsiders in this place that was rightfully theirs.[3]

Mabel's remarks brought home to me the extent to which country, for Aboriginal people, is a *social* reality, steeped in memories of births, deaths and marriages, of seasonal movements in search of food, and of the traumatic disruptions of colonial history. But it is not passively being on or in the land that gives the land its vitality and meaning; nor are these qualities the result of contemplation. Rather, it is the *vita activa*, the process of living and moving with others on the land and drawing one's livelihood from it, that charges the landscape with vitality and presence. Country embodies the sweat, energy, thought, and feelings of those who invest their labor in it, just as a fabricated object becomes charged with the vitality of the person who shapes it. Like the Ionian theorists of nature in the sixth and seventh centuries, Aboriginal people assume the world of nature to be "saturated or permeated by mind."[4] The ebb and flow of tides, the fury of storms and earthquakes, leaves buffeted and trees broken by high winds, all testify to the ways in which nature is not only filled with energy and power, but "ensouled."[5] Accordingly, relationships between realms that we conventionally separate as natural, cultural, and supernatural are all glossed as *social* relationships, governed by the same principles that obtain in interpersonal life. Among the Kuku-Yalanji, such analogical reasoning means that the ecological zones of "sea" and "inland" are also cultural categories—"of the sea" *(jalunji)* and "away from the sea" *(ngalkalji)* connote separate moieties whose members have different essences and may be identified by their different smells. This logic also explains why sea and inland things must be kept apart. So one is enjoined not to use dugong, turtle, or bullock (which are "meat") as bait for fishing, but to use only fish bait to catch fish (the others being "whitefella bait"). And don't use saltwater fish to catch freshwater fish, or vice versa, one is told. To infringe any of these *cultural* rules will cause a flood . . . an ungovernable overflowing of *natural* boundaries.

Thus one learns that misfortunes that would in one's own life-world be dismissed as accidents, or regarded as simply in the "nature of things" actually have social causes; *someone* must be responsible for them, *someone* must be to blame. The same reasoning explains why "natural" phenomena are so closely and continually examined for their *social* implications, as when a shooting star or a kookaburra laughing at first light are said to signal a death.

This is not to imply that Aboriginal and Western worldviews imply absolutely different life-worlds. That these worldviews seem so incommensurate may be more an artefact of our longstanding habit of exoticizing "primitive" people than a reflection of any empirical reality—a habit still evident in the tendency of many contemporary philosophers of ecology to excoriate global capitalism by urging a recovery of the allegedly more eco-sensitive, sensuous, reciprocal relation between humanity and the natural world that pre-modern thought is said to epitomize. All such constructions of the other are deeply flawed. In the first place, they inevitably construct nature as benign, and narcissistically invoke experiences of the natural world that are pleasing rather than destructive or discomforting to us. The Kuku-Yalanji notion of storms as the malevolent expression of human ill-will, of lightning as retributive justice, and of earthquakes and volcanoes as signs of the earth's outrage, call such romanticism into question. In the second place, such constructions gloss over the fact that a sensuous experience of connectedness between people and their environment is never permanent or pervasive, but always occasional—arising in specific social contexts, tied to specific social purposes, and constrained by cultural ideas and ritual codes. That Mabel Olbar and McGinty Salt, our hosts, made keen observations of the bay whenever we arrived there to fish—remarking the spoor of a snake in the sand, traces of mullet or herring offshore, the state of the tide, and subtle nuances of the sea, the weather and the season that entirely escaped my notice—was not because they *participated* in nature but because they were *practiced* in that way of life in that place.

In this sense their participation in the place they called their own was no more "mystical" than the participation of a mechanic, say, in an assemblage of machine parts on which he is working, or a scholar in an engrossing text, or a sculptor in the object she is shaping. All, so to speak, put themselves into what they do, creating thereby the conditions under which they may experience that sense of fusion between body-self and object that we tend to talk about in terms of naturalness, sympathy, and attunement. In short, states of consciousness, as Marx repeatedly observed, are tied to our modes of *interaction* with the world in which we live.

Two days into our stay, and after long hours working with McGinty and his brother-in-law Babaji to set up our campsite, I went down to the bay alone, stripped, washed and scrubbed myself in the sea, then dressed. The beach was deserted. But as I sat in the shade of a pandanus palm, thinking back on the day, and on the fulfilment I had found clearing our campsite with McGinty and Babaji, an aluminium dinghy came slowly inshore from the open sea.

It was Mabel's brothers, Sonny and Oscar, and her brother-in-law Sam. They had been out to Hope Island, hunting green turtles. As they beached the dinghy and drew it up onto the sand, I went down to greet them.

The sea sloshed around my ankles and gently jolted the dinghy. The two boys, Philip and Louie, ran down the beach brandishing their fishing spears as Sam and Sonny hauled the biggest turtle onto the gunwale of the dinghy and tied a rope around its right flipper. Then, as the old man of the sea appeared to gaze about, befuddled, Sam beat its brains out with a sledge-hammer.

I watched intently as Sonny began to butcher the turtle.

"We call turtle 'meat' *(minya)*, not 'fish' *(kuyu)*," Sam explained. And he told me that great care had to be taken when separating the meat from the carapace, for if the bile is spilled it contaminates the meat and makes it inedible.

In the face of such pragmatism, what place did my own unspoken sentiments have, as I watched this beautiful creature—so out of its depth, so out of its element—being hacked open before my eyes? And how might one reconcile the great difference between the Aboriginal and non-Aboriginal sensibilities that collect around such an event? For while many eco-conscious Australians regard the green turtle *(Chelonia mydas)* as both a beautiful and endangered species, Kuku-Yalanji regard its green fat as a delicacy, and hunt and eat it with relish.

That evening, Sonny disinterred the cooked turtle from the earth oven he had dug at the outstation, and the fat was shared around. I ate without much appetite, caught between competing cultural persuasions.

Green Turtle

Sam smashes its head in
with the same sledge-hammer
I used this afternoon
to ram our tent pegs home.

A hemisphere turns
turtle; Sonny hacks
its mildewed, sea-marbled
breastplate free.

It recoils from the sky.
Head lolls.
A flipper feebly pushes
Sonny's knife away.

They empty
the long grey rope of its life
onto the sand by the thudding boat
which holds two more

And its carapace is a vessel
filled with a wine lake
in which clouds
float, birds fly, leaves fall.[6]

As the months wore on, I came to understand how Kuku-Yalanji read their environment, learning, for instance, that a hammer bird heard in the cold months means that mullet will be plentiful, that bean trees flowering or the wild tamarind ripening mean that scrub hen eggs can be found, and that the flesh of the parcel apple turning pink means that the liver of the stingrays will also be pink and therefore good to eat, though eating stingrays in the preceding months (October-November) will bring storms. As the wet season approached, I became increasingly fascinated by the family's preoccupation with thunder and lightning. Whereas I saw storms as natural phenomena, our hosts interpreted them in social terms; they were expressions of human malevolence and of tempestuous states of mind. Thus, the phrase *jarramali bajaku* (literally, "exceedingly stormy") is used of persons who lose self-control when drunk or drugged, while the term *jarramali* denotes a cyclonic or monsoonal storm, any one of which may embody the ill-will of outsiders. Questions of control thus entail allusions to individual psychology, relations with others, and relations with the elements of nature. Put another way, the "environment" includes social beings, asocial beings, natural species, natural phenomena, and innate "natures."

In Aboriginal communities, one is often struck by people's extraordinary tolerance of aberrant or unruly behavior. I was sometimes reminded of my experience among the Kuranko in Sierra Leone where incorrigible individuals would draw such comments as, "He came out of the *fafei* like that" (i.e. even initiation failed to make him mend his ways), or "That is how he was made" *(a danye le wo la)* or "He is blameless; he was born with it" *(a ka tala; a soron ta la bole)*. But while both Kuranko and Kuku-Yalanji explain dispositions that resist socialization by invoking notions of innateness, there are practical limits to people's tolerance of antisocial behavior which, in both societies, is seen as a form of deafness to social values.

It was Christmas 1993. The heat and humidity was oppressive. Sweat dripped from my forehead onto the pages of my journal as I wrote about the tension that had built up in our camp, breaking on Boxing Day (the day after Christmas) like a storm, with Sonny in a fist fight with his brother-in-law, his elder sister heaping abuse on his head, another sister throwing a couple of punches for good measure, and then the youngest sister Gladys and her husband driving off to Ayton to get away from it all. As the first thunderstorm of the wet season approached, the sky turned indigo and the wind veered and picked up. There was a rattle of dry leaves and dry leaves falling, for which Kuku-Yalanji use the word *yanja*, followed by the crumpling sound of distant thunder, like heavy furniture being moved around in an upstairs room—a sound that also has its own specific ideophone, *kubun-kubun*. Painstakingly, people tracked the course of the storm, discussing where it was coming from and where heading, identifying its sounds, observing its effect on the foliage, comparing it with storms in the past. Indeed, the character of the impending storm was analyzed in the same way that people analyzed strangers—trying to read their intentions, second-guess their motives, identify their mood. As this discussion went on, various members of the family made forays into the bush in search of wild grape *(kangka)* vines, ironwood bark *(jujabala)*, and grass tree *(nganjirr)*. Sonny, now sober, applied himself to the business at hand, burning knotted hanks of grass, ironwood bark, and grass tree outside our camp. As the sweet smell of the grass tree pitch *(kanunjul)* spread across the clearing, I assumed that it was meant to repel mosquitoes. But Sonny told me that the storm would smell the smoke and go away. I later asked McGinty, who was not Kuku-Yalanji, if he could explain to me how burning grass tree could ward off storms. The idea seemed to both amuse and embarrass him, partly because his own people on Princess Charlotte Bay used a different method of warding off storms (a certain kind of shell), partly because he did not want to give me the impression that he was a superstitious *myal* ("bush person"). That evening, as I was helping him put up his tarpaulin and tent at the beach, he joked about the ominous rain clouds hovering over the range. "Might rain soon," he said laconically. "Better tell that storm to wait until I get my tent up."

At Mabel's sister's house in Ayton, however, the mood was somber. Most of the family had gathered behind closed doors, huddled and anxious as the storm approached. One of the children gave my wife a clue as to why they were so fearful: "If you eat things you are not supposed to eat, a storm will come and punish you." Was lightning an agent of retributive justice, seeking out those who might have broken a food or sex taboo, or transgressed a sacred site? Such matters are difficult for any anthropologist to divine, for who knows what guilty secrets a person may harbor, and whether these get projected as fears of external retribution. One thing was clear, however, and that was the association of thunderstorms and vengeful outsiders.

In the 1890s, the ethnographer W. E. Roth reported that in many parts of northern Queensland, thunder and lightning were means of sorcery, but that people sometimes summoned these same forces to drive white settlers from their land.[7] I heard identical stories from my Kuku-Yalanji friend, Harry Shipton, in 1993. Many years ago, Harry told me, a white rancher, exasperated by *bama* (Aboriginal people) spearing his cattle, rode up to a river encampment and shot a young girl dead. Bent on revenge, the girl's father went to the rancher's place as thunder. The rancher fired shots at the thunder but his bullets passed harmlessly through the thunder's body. Then, with a single lightning bolt, the thunder speared and killed the rancher. In another of Harry's stories, a certain white man who "messed with many *bama* girls," getting them pregnant and causing trouble, was sought out by lightning as he was driving his tractor in a Mossman cane field. "Bang! he dead, just like that."

The association of storms with sexual desire, jealousy, and revenge was further clarified for me by Sonny Olbar. One day, I observed Sonny knotting hanks of grass and stuffing them under some logs of grass tree *(Xanthorrhoea arborea)* and ironwood bark before setting fire to them "to keep the thunderstorm *(jarramali)* away." When I asked him to explain, Sonny said, "Thc storm smells the *nganjirr* (grass tree) and goes away."

After more questioning, I figured out that the underlying logic here rested on an analogy drawn between one's relations with in-laws (who, because of the rule of exogamy, are comparative strangers) and one's relations with thunderstorms (that also come from elsewhere).

The key terms, and the relationships between them may be posited thus:

Mother-in-law : son-in-law :: thunderstorm : grass tree.

When thunderstorms approach, it is feared that social categories that should be kept apart are coming dangerously close together: oneself and one's enemies, insiders and outsiders. This situation is compared to the infringement of the avoidance relation between mother-in-law and son-in-law, and by association any transgression of things that should be kept apart, such as people and forbidden fruits.

The problem: how to drive the thunderstorm away?

The solution: activate the analogies alluded to above.

The practical action: grass tree logs are burned.

The explanation: grass tree (as well as iron tree bark and wild grape vine) is son-in-law to the thunderstorm. The thunderstorm will smell the grass tree smoke. And just as a mother-in-law will avoid her son-in-law if she smells him, so the storm will move away when it gets wind of its son-in-law, the grass tree.

This brief excursion into Kuku-Yalanji ethnography enables us to see that the wild powers of what we call nature are metaphorically fused with environmental forces that we call social and political—the world of white-fellas and the Australian state, the world of cultural outsiders and of affines. These external environments offer a wealth of possibilities for improving one's standard of living—imported commodities, social services, government grants, family networks, outsiders like anthropologists with useful expertise. But gaining access to such life-enhancing resources involves dealings with strangers that one can never fully understand, trust, or control, and the external environment remains a mixed blessing, a place of both positive potentiality and invisible dangers.

The Kuku-Yalanji land claim has been settled with the Kuku-Yalanji establishing their rights over much of the area, although there are large national park areas in which they will be sharing the management of their traditional rainforest and coastal environments with the Australian State. But it is perhaps worth reminding ourselves that there is no landscape, no ocean, and now no sky, that has not been changed irrevocably by the work of human hands and the human imagination. When James Cook sailed along the coast of southeast Cape York in June 1770 after his ship had been holed on the barrier reef and his crew did not know whether they would ever see their homes or loved ones again, he observed unforested hills where now there is rainforest that one assumes to be primeval and virgin. It is hard to know how the landscape will judge us years hence—we who hold such radically different views of it, each one of which seems imperatively true to the believer, who is certain he or she knows what things were like in the past, what the future will bring, and who deserves to inherit the earth.

NOTES

1. Kay Milton, *Environmentalism and Cultural Theory: Exploring the Role of Anthropology in Environmental Discourse* (London, New York: Routledge, 1996), 109–114.
2. Christopher Anderson, "Aborigines and Conservation: The Daintree-Bloomfield Road," *Australian Journal of Social Issues* 24 (3) (1989): 214–227.
3. The material from"This is my *buba*, my country," to the end of the poem, *Green Turtle*, appeared in a slightly different form in my memoir, *The Accidental Anthropologist*. See Michael Jackson, *The Accidental Anthropologist* (Dunedin, New Zealand: Longacre Press, 2006), 301–305, used with permission of the author and Longacre Press.
4. R. G. Collingwood, *The Idea of Nature* (Westport, Connecticut: Greenwood Press, 1944), 3.
5. Ibid.
6. "Green Turtle" first appeared in Michael Jackson, *Antipodes* (Auckland, New Zealand: Auckland University Press, 1996), 9. Used by permission of the author and the press.
7. See Walter Roth's work, particularly, Walter E. Roth, *Ethnological Studies among the North-West-Central Queensland Aborigines*, (Brisbane, Australia: Edmund Gregory, 1897), 168 and Walter E. Roth, *North Queensland Ethnography: Superstition, Magic, and Medicine* (Home Secretary's Department, Brisbane, Bulletin No. 5. Brisbane: G.A. Vaughan, Govt. Printer, 1903), 8.

Notes on Contributors

Donald K. Swearer is Distinguished Visiting Professor of Buddhist Studies at HDS and serves as Director of the Center for the Study of World Religions. Before coming to HDS in 2004, Professor Swearer taught at Swarthmore College as the Charles and Harriet Cox McDowell Professor of Religion. His main research areas are: Theravada Buddhism in Southeast Asia, primarily in Thailand, and Buddhist Social Ethics. His current research interests focus on sacred mountain traditions in northern Thailand as well as Christian identity in Buddhist Thailand. His recent books include *The Buddhist World of Southeast Asia* (the second revised edition in press); *The Legend of Queen Cama: Bodhiramsi's Camadevivamsa, a Translation and Commentary* (1998), *Sacred Mountains of Northern Thailand and Their Legends* (2004), and *Becoming the Buddha: The Ritual of Image Consecration in Thailand* (2004). Among his recent courses are Buddhism, Ecology and the Sacred Mountain Traditions of Asia; The Buddha in Myth, Legend, and Ritual; Religious Belief and Moral Action; and Buddhist Social Ethics. Professor Swearer has published several essays on Buddhism and ecology and is a founding board member of the Forum on Religion and Ecology.

Lawrence Buell is Powell M. Cabot Professor of American Literature at Harvard University, where he has taught since 1990. His research interests include rethinking U.S. literature in a globalizing world, discourses of literature and environment, and the theory of national fiction. He is the author, among other books, of *New England Literary Culture* (1986), *The Environmental Imagination: Thoreau, Nature Writing, and the Formation* *of American Culture* (1995), *Writing for an Endangered World: Literature, Culture, and Environment in the United States and Beyond* (2001), *Emerson (2003), and The Future of Environmental Criticism* (2005). *Writing for an Endangered World* won the Popular Culture and American Culture Associations' Cawelti Prize for the best book of 2001 in the field of American cultural studies; *Emerson* won the 2003 Christian Gauss and Warren-Brooks prizes for outstanding literary criticism. In 2007, the Modern Language Association honored Professor Buell with the Jay Hubbell Award for lifetime contributions to American literary studies.

Michael D. Jackson is an anthropologist, and has carried out ethnographic fieldwork in Sierra Leone and Aboriginal Australia. The author of numerous books of anthropology, including the prize-winning *At Home in the World* (1995), he has also published three novels and six books of poetry. His most recent books are *Excursions* and *The Palm at the End of the Mind*. Michael Jackson's work has been strongly influenced by critical theory, American pragmatism, and existential-phenomenological thought. In his ethnographies he has sought to show how reflection and research can engage with the everyday issues, exigencies and struggles that characterize human life in every society, irrespective of their historical and cultural differences. Michael Jackson has taught in his native New Zealand, Australia, the United States (Indiana University), and Denmark (University of Copenhagen). Since 2005, he has been teaching at Harvard Divinity School as Distinguished Visiting Professor in World Religions.

Daniel P. Schrag is Professor of Earth and Planetary Sciences, Harvard University, and Director of the Harvard University Center for the Environment. He studies climate and climate change over the broadest range of Earth history. A former MacArthur Fellow, he is widely published in scientific journals. He has examined changes in ocean circulation over the last several decades, with particular attention to El Niño and the tropical Pacific. He has worked on theories for Pleistocene ice-age cycles including a better determination of ocean temperatures during the Last Glacial Maximum, 20,000 years ago. Schrag also helped develop the Snowball Earth hypothesis, proposing that a series of global glaciations occurred between 750 and 580 million years ago that may have led to the evolution of multicellular animals. Currently he is working with economists and engineers on technological approaches to mitigating future climate change.

Since **Bron Taylor**'s appointment as the Samuel S. Hill Ethics Professor at the University of Florida in 2002, he has taught in the world's first graduate program focusing on religion and nature. He has written widely about environmental movements and environmentalism globally, with special attention to their religious, moral, and political dimensions. He is president of the International Society for the Study of Religion, Nature and Culture and editor of the *Journal for the Study of Religion, Nature and Culture*. His works include *Dark Green Religion* (appearing in 2009), the award-winning *Encyclopedia of Religion and Nature* (2005), and *Ecological Resistance Movements: The Global Emergence of Radical and Popular Environmentalism* (1995). He also hosts the website, www.religionandnature.com, which acts a gateway to his initiatives, research, and teaching.

Mary Evelyn Tucker is codirector of the Forum on Religion and Ecology (www.fore.research.yale.edu). She is the author of *The Philosophy of Qi* (2007), *Worldly Wonder: Religions Enter Their Ecological Phase* (2003), and *Moral and Spiritual Cultivation in Japanese Neo-Confucianism* (1989). She also edited two volumes on Confucian spirituality with Tu Weiming and the CSWR's ten-volume Religions of the World and Ecology series with John Grim. She received her PhD from Columbia University in East Asian religions. Until 2005 she was a professor of religion at Bucknell University, including serving as a National Endowment for the Humanities chair from 1993 to 1996. She is currently a Senior Lecturer and Research Scholar at Yale University with joint appointments in the School of Forestry and Environmental Studies, the Divinity School and the Department of Religious Studies. She also is a research associate of Harvard's Reischauer Institute of Japanese Studies.

Donald Worster is Joyce and Elizabeth Hall Professor of History at the University of Kansas. He received a BA in 1963 and an MA in 1964 from the University of Kansas. He continued his education at Yale University, earning an MPhil in 1970 and a PhD in 1971. Dr. Worster's research, lecturing, and teaching fields include the environmental history of North America and the world, the American West, and nineteenth- and twentieth-century U.S. history. He has held teaching appointments at Brandeis University, the University of Hawaii, and the University of Maine. He serves on the boards of several environmental organizations. His publications include *A River Running West: The Life of John Wesley Powell* (2001), *An Unsettled Country: Changing Landscapes of the American West* (1994), *The Wealth of Nature: Environmental History and the Ecological Imagination*

(1993), and *Nature's Economy: A History of Ecological Ideas* (1977). His most recent work is *A Passion for Nature: The Life of John Muir*, published by Oxford University Press in 2008.

Michael Zimmerman is Professor of Philosophy and Director of the Center for Humanities and the Arts at the University of Colorado, Boulder. Until 2006, he was chair and Professor of Philosophy at Tulane University and codirected Tulane's Environmental Studies Program for a decade. His academic specialties include recent German philosophy and environmental philosophy. Author of four books, *Eclipse of the Self: The Development of Heidegger's Concept of Authenticity* (1981), *Heidegger's Confrontation with Modernity* (1990), *Contesting Earth's Future: Radical Ecology and Postmodernity* (1994), and the forthcoming *Integral Ecology: Uniting Multiple Perspectives on the Natural World* (co-authored with Sean Esbjorn-Hargens), Zimmerman is also general editor of a leading anthology, *Environmental Philosophy: From Animal Rights to Radical Ecology* (1993). Zimmerman has also published almost one hundred articles and book chapters.

Index